How To Keep Hens For Profit

by C.S. Valentine

with an introduction by Jackson Chambers

This work contains material that was originally published in 1913.

This publication is within the Public Domain.

This edition is reprinted for educational purposes
and in accordance with all applicable Federal Laws.

Introduction Copyright 2017 by Jackson Chambers

Self Reliance Books

Get more historic titles on animal and stock breeding, gardening and old fashioned skills by visiting us at:

http://selfreliancebooks.blogspot.com/

Introduction

I am pleased to present yet another title on Poultry.

The work is in the Public Domain and is re-printed here in accordance with Federal Laws.

As with all reprinted books of this age that are intended to perfectly reproduce the original edition, considerable pains and effort had to be undertaken to correct fading and sometimes outright damage to existing proofs of this title. At times, this task is quite monumental, requiring an almost total "rebuilding" of some pages from digital proofs of multiple copies. Despite this, imperfections still sometimes exist in the final proof and may detract from the visual appearance of the text.

I hope you enjoy reading this book as much as I enjoyed making it available to readers again.

Jackson Chambers

PREFACE

A CONSIDERABLE portion of the material here presented was first published in the columns of the *New York Tribune Farmer*. Permission has been kindly granted the author to use this material in book form, and to it has been added about one-third new matter to round out the subject under discussion and bring it down to date. It is hoped that this first presentation of the American class of fowls as a group meriting an entire book to itself may find favor with the many who know the real "American" hen, and the more who ought to know her.

CONTENTS

	PAGE
THE NATION AND THE POULTRY INDUSTRY	1
THE PLYMOUTH ROCK, THE JAVA, AND THE DOMINIQUE	10
THE WYANDOTTE	23
THE RHODE ISLAND RED AND THE BUCKEYE	36
CHIEF COMPETITORS OF THE AMERICAN HEN	52
THE FIVE-DOLLAR-A-YEAR HEN	60
IMPROVING THE AMERICAN HEN	68
QUALITY AND NUMBERS AT THE SHOWS	79
THE AMERICAN FARMER, THE AMERICAN HEN, AND THE FANCY	86
THE AMERICAN HEN AND AMERICAN MONEY	110
MAKING A HIGHLY PROFITABLE WORKER	124
EGGS OF THE AMERICAN BREEDS	141
AVERAGE LAYERS AND THREE-HUNDRED-EGG HENS	155
EGG FOODS: CAN THEY INSURE US A TWO-HUNDRED-AND-FIFTY-EGG HEN?	167
MOTHER AND CHICKS	180
HANDLING THE CHICKS	188
MODERN WAYS OF HOUSING	211
EXPENSIVE ACCIDENTS	238
COMMON-SENSE HANDLING OF COMMON DISEASES	249
THE INDIAN RUNNER DUCK	280
INDEX	293

How To Keep Hens For Profit

THE NATION AND THE POULTRY INDUSTRY

THE advance sheets of the United States Government report for 1909 are enough to stir the blood of any patriot. "But this is prosaic," you say? "this is far from being patriotic." May we differ? The patriot is not only one who has sentiment for his country, but who "supports its authority and its interests." I submit that the greatest patriot of all is the farmer; for it is he, year by year, who most consistently supports the interests of the country — nay, who saves its very life, individual and collective.

One of the first affirmations of the advance sheets above noted is that the value of the farm products is now so almost incredibly great that it "has become merely a row of figures." The gain in values, over last year, is eight hundred and sixty-nine millions of dollars; while, so short a time as ten years ago, the entire production was only five and one-half times this present gain. With such strides are we advancing!

To-day, in Washington, lives a man, interested in poultry raising from many points of view, who likes to tell how he once had to labor with the heads of Departments in Washington to get any recognition of the high place which poultry was even then taking among the products of this country. Up to that time poultry had been largely ignored.

The commanding place which poultry has now taken by right of seizure is being universally admitted. The Government devotes both money and men to its interests, and gives it prominent space and appreciative notice in the annual reports. It emphasizes the fact that the poultry crop is now equal to the hay, the wheat, or the cotton crop, only corn having greater value than these. It published between 1896 and 1907 no less than eleven new, special bulletins having to do with poultry, besides modernizing some others, and has devoted more than fifty pages to poultry, in a single year's report. It has established an experimental poultry plant of its own near the capital. It has conducted investigations in poultry diseases and in cold storage of poultry and poultry products.

During this latter investigation it placed an observer in one of the commercial cold-storage plants in one of the largest cities. This worker, finding that the very important instruments for the determination and regulation of the moisture content of the atmosphere then available were not giving satisfactory service, devised one which met the need more thoroughly. The Government applied for a patent on this device, in order that any one in the United States might use it free of royalty. It has also made a study of markets and transportation as affecting poultry products, especially eggs for market. In all this the Government is but conserving its own strength, since this depends on the welfare of the farmers, the great supporters of the poultry industry.

The United States Government gives also a notable portion of its efforts to the preservation of the wild bird life within its jurisdiction. Game, strictly speaking, does not come within the limits of the present study. But we may take a little time to look at the present conditions of game supply and game protection, because the supply of game affects, to some degree, the poultry market. And, through the recent extension of the culture of a duck which is decidedly "gamy" as to meat, game supply becomes a question of interest to many farmers. This is especially the case because this is the best layer ever known among domestic ducks, it having just been reported as making a phenomenal average record, in an officially handled competition, and an actual individual record therein of three hundred and twenty eggs.

The price of game has been so high that the fact

of restaurant portions being billed on the menus in terms of dollars instead of cents has excited little comment. The average wholesale price of dressed poultry in New York City was a trifle over fifteen and one-half cents a pound, in 1908. But the guinea-fowl, which may almost be classed as half wild, though in domestication, has crept into the opening made by the scarcity of game, and thus advanced in price in unusual ratio. A prominent commission house wrote me, some months ago, that guinea broilers, especially, were in firm demand at a very good price.

The failure of the available game to supply the demand for game has drawn the attention of the Government to its protection, preservation, and increase. The number of hunters annually in the open is now estimated at above three millions. The capture of birds for millinery purposes is under some control; but civilization is constantly encroaching on the breeding haunts of the birds. Among quail in covers a disease has appeared, which discouraged increase of this interest.

As far back as 1905, thirty-six states had state officers for the protection of game, and non-resident hunting licenses were required in all, while about half required resident hunters also to be licensed. There has been so much hunting in this country by non-resident foreigners, that North Carolina alone, in one year, received ten thousand dollars for such licensing. In other localities, in one case, four hunters were fined one hundred and forty dollars for killing six meadow-larks; a fine of four hundred dollars was enforced for shipping two hundred prairie chickens and other birds. In still another case, two hundred dollars and costs were imposed for shipping twenty quail. For possession and sale of quail and grouse in the close season, New York has exacted several fines ranging from three hundred dollars to six hundred dollars each.

Near the middle of the decade following 1900, came one or two winters of extreme severity, when quail suffered intensely. So many died that some states were driven to restocking. Massachusetts could not get birds at first; New Jersey fared better, and liberated more than eight thousand. The suffering and loss during these two very cold winters caused a systematic effort to have game birds fed. In one state the state zoölogist sent out thirty thousand requests to farmers to feed the birds. In Massachusetts the Protective Association distributed cards, urging people to feed quail, furnishing feed free of charge, with instructions how to place it so that it would do the most good. In Illinois the sum of three thousand three hundred dollars was spent for feeding game birds during the severe months. Besides the efforts toward preservation, importations have been made of quail, partridges, and grouse, from Mexico and Europe.

In 1905 the President of the United States was authorized to set aside, at his will, portions of the Wichita Forest Reserve, in Oklahoma, as a game sanctuary. In June of that year he designated fifty-seven thousand one hundred and twenty acres as a preserve, this being especially adapted to the propagation of wild turkeys and quail, as well as of some larger game.

Illinois leased one hundred and sixty acres for ten years for the special propagation of pheasants, quail, and grouse. In Pennsylvania the Board of Game Commissioners was authorized to establish public preserves in the State Forest Reservations, for the increase of wild turkey, partridge, quail, and woodcock, as well as wild pigeons. In Maine the Audubon societies have leased one island, said to be the only known breeding place of the eider duck in this country.

Within the year immediately following August, 1907, the President set aside nine new bird reservations, and during 1909, the increase in the number of such reservations was from sixteen to fifty-one, all on land of little or no agricultural value. Two are on marsh land; while one embraces several of the Hawaiian Islands, and United States artificial reclamation lakes form others. The Government is also stocking ponds and waterways with wild rice, wild celery, and other favorite foods for the water fowl. In this way it is hoped to increase the number of wild ducks and geese. For propagation, as well as preservation, nearly thirty thousand gray partridges and twelve hundred pheasants were imported for liberation in the United States, and over five thousand eggs of pheasants, partridges, and wild ducks have been imported, all during 1909.

The Government officials believe that each year brings a better understanding of the need for game protection. It is still true, according to my own observation, that farmers look upon hunting and fishing for needed food as a hereditary right belonging to the land and the owners thereof, despite forbidding game laws. They regard huntsmen from cities and clubs, in the mass — especially if mounted — as being quite as predatory as were the bands of cattle thieves which overran the Scottish borders in the older days. They find it difficult to believe that the pheasant which destroys their crops with yearly increasing impudence ought to be protected; but they are awakening to the fact that the non-game birds are their neglected friends — and ought to be their intimates — and are taking a live interest in their preservation.

All food meats are bound together in a community of interest, in that the failure of any important one, or of several not so important, may have a strong bearing on the future of any others which might take the places of these former supplies. Investigations in fifty cities have shown that retail prices on meats have a varying range of gross profits from 17 per cent up to 68 per cent above wholesale prices. The south central group of states shows a mean profit of 54 per cent above wholesale prices. I think there has never been known such a profit on eggs. But so long as conditions combine to permit such profit on cattle meats, the prices of eggs will go up in sympathy, for many more eggs will be used when meats are high in price than when they are low. High prices for feed also push in the same direction. The per capita consumption of meat for all the people of the country, the very young as well as the adults, was, for 1900, 182.6 pounds. This demands billions of pounds for supply.

"The price of corn has been too high for the price of pork," says the statistician. That the same is true of corn as compared with eggs, in price, is the testimony of every poultry raiser. But eggs in 1908 have averaged more than twice as high as the average of the four years preceding the census of 1900. Farm products as a whole averaged but 26 per cent higher, so that eggs are decidedly in the lead as a source of income for those who can produce them judiciously. The decline in the supply of beef and other animals means the further advance in prices of eggs and poultry, unless it were possible that the systematic efforts to conserve and increase the supply of game should be — shall I say *unduly?* — successful. Should these be so widely successful as to increase the supply of game meats sufficiently to affect the general meat market, we might wonder what would be the outcome. But I think no one would consider this possibility seriously at the present time.

There is still another important point of view: only such products as can be gained through domestication of any subject lend themselves easily to extreme aggregations of commercial values. Admitting that their quality suits the market, the Guinea-fowl and the Indian Runner Duck are more desirable producers of "game" meat than are the wild fowl. This principle the race has virtually admitted, ever since it began to domesticate the wild animals and the wild fowls. It is this which paved the way for the present well-nigh marvellous performances of the American Hen. It is this which may lead to unheard-of successes in some of her near of kin, such as the Guinea-fowl and the Indian Runner Duck.

THE PLYMOUTH ROCK, THE JAVA, AND THE DOMINIQUE

Wide Distribution of the Rock — The Egg Type — Large Claims and a Poor Showing — A Find — Handicaps for One-color Sorts — Troubles of the Buffs — Care in Feeding — Muscular Development — The Java-Dominique Rock — The Good Old Java — Why does not the Dominique Score?

It is now about twenty-five years since the then best-informed man in the country, probably, on this subject made a crucial assertion, and one that showed his Yankee extraction. He said that the Plymouth Rocks and the Wyandottes were two of the four best-paying breeds known to him, from the practical side. Even at this early day the American breeds were proving their right to have been created for Americans: they paid! This speaker referred to the Barred Plymouth Rock and the Silver Wyandotte, as we now have to term them because of the several varieties which each breed now boasts.

The "cuckoo" plumage of the Barred Rock has always been widely liked, and the fact that the birds do well in average hands has made them favorites on the farms, where everything else often has to give way to the exigencies of growing and gathering the grain crops.

It is one thing to read in some irresponsible print that hundreds of millions of dollars for the farmer lie, embryonic, in the poultry industry. It is quite another thing to have the Government officials of this great country stand sponsors to such broad statements as have been noted in the previous chapter. And if any one breed of fowls might claim the palm of being the largest influence in forcing such an official admission, the Plymouth Rock, especially in the Barred variety, may lay a reasonable claim to being that one.

The Barred Plymouth Rock, the kind that used to stand for the name "Plymouth Rock," years before there was any other variety, is probably the best-known bird in this country. It is also claimed that it is the most widely raised. Perhaps the one proves the other. But I wish to call attention to the fact that this particular marking is very persistent, and that it comes out in birds with only a fraction of the Plymouth Rock blood, so that much of the claim for the Rock's wide distribution on the farms embraces birds that may best be named mongrels, with more or less of Barred Rock blood. The fact that color foreign to the true Barred Rock Standard requirement is so prevalent in these farm birds is perhaps sufficient proof of the above statement. "Dirty-looking birds," in the thought that their color is not clear, the great majority of them are pretty sure to be.

The true, highest-class, show-room Barred Rock is a bird for breeders to go into raptures over, especially those who know how very difficult it is to produce. But it is probably wholly true that no novice can hope to go into the open market and buy stock that will give him birds of the highest class, unless he stumble into the hands of a good breeder, who will coach him on the manner of handling the double-mating system, and who will sell him the right matings to start with.

The ordinary buyer, ignorant of the ways in which strains are produced, and of the double-mating used to produce winning Barred Rocks in many instances, could do no worse than to buy from one of these established strains, unless he put himself into the hands of their breeder for instruction, and for new birds when he needed them. If he first got birds from double-mating, he must know exactly how to handle them. Should he mate them with outside blood, he would get a hodge-podge, probably all culls, from the big exhibitor's point of view. It is never safe to breed Barred Rocks without knowing what is back of them in blood and handling, and the new breeder has so much to learn that it may well make him dizzy to contemplate it. If he likes hard knots, he will have fun; if he isn't fond of puzzles, it will be anything but fun, unless he has a "coach," and one that stands by him.

It often seems strange to me that the Barred Rock, an American breed, and best known among farmers at large of all the breeds, should be so little known among them as to its fancy points. I am not sure that it is a very good breed, however, for a farmer to choose when he wants to breed to fancy points, because it is one of the most difficult breeds known to bring to perfection, or anywhere near that desired quality. And when it comes to showing, there are so many old breeders who raise high quality Barred Rocks that it is well-nigh impossible for a novice to win anything worth while in the large shows.

Color and its proper distribution count for well toward half the value of an exhibition Barred Rock. More than forty points out of the one hundred that mean absolute perfection are allowed for color, from three to six points being counted for color in each section at the present time. Indeed, this is the scale given for all breeds in the American class, which includes four varieties of Rocks, eight varieties of Wyandottes, two of Javas, one each of Dominiques and Buckeyes, and two of Rhode Island Reds. Shape has thirty-three points, comb eight, symmetry and condition twelve more. Weight counts six points, and the specimen loses on each half pound under weight, the bird nearest

Standard weight gaining the prize, if other points are equal. The Standard weight of Rocks runs from six and a half pounds in pullets to nine and a half pounds for mature cocks. Hens are one pound heavier than pullets. The emphasis which breeders put on head points may be understood if we know that the head of a bird in the American class counts twenty points, including shape, color, comb, wattles, and ear lobes. In the Mediterranean classes it counts six points more; that is, more than one-fourth the value of the bird. At first thought, one might be led to think scornfully of fancy fowls when considering this requisite. But it is also true that the head is a very good index to the practical value of the bird, though not all the points count in this grading.

The color of the Barred Rock has been a matter of contention and misunderstanding for many years. The best that the combined judgment of breeders can do even now in describing it is to call it a modified black and white, adding a limiting clause to the effect that overlapping of feathers and reflection of light cause the feathers to show a bluish tinge. Some still contend, I think, that the color should be described as in shades of blue or blue-black. The perfect specimen should be barred to the skin and be even in surface colors throughout.

In exhibition fowls, generally, the fluff, the folding of

the wings, the setting of the legs, the curves and angles of the body, the texture of comb and wattles and lobes, all form a part of the symmetrical whole which approaches perfection, yet never quite reaches it.

It is significant, I think, that the American Standard of Perfection gives Barred Plymouth Rocks first place in its pages, after the preliminaries. The birds nearest to Standard weights are declared to be not only most typical, but to meet the utility requirement best, as well. The three varieties are asserted to be identical, save as to color. The color nomenclature of the Barred Rock has been a matter of fighting of the kind that "draws blood," and most breeders admit that it is a color one must learn to know by studying the best birds. Neither "blue" nor "black" used alone gives the right idea of this "shifting shade."

The show-room Barred Rock is not the "rock" on which the country stands, however, but rather that approach toward her which is generally known by the breed name. I think there is no doubt that a farm fowl which is party-colored is in many senses more desirable than a fowl having one plain color, especially if that one color is light. The farm breeds poultry, first of all, for its own use, and it wants a type that can be raised easily in large numbers and that does not require too nice attention. This is speaking broadly; of course, many farmers are breeding to Standard, and giving close attention to fine points, but they count small beside the great majority who want, first of all, a utility fowl.

I once heard a student ask an expert what constituted the egg type in his particular breed. The answer was: "The bird nearest to Standard requirements." And this was from an old man who had spent a lifetime in breeding fancy fowls. I think nearly all fanciers would make the same claim, but it does not seem to me to be wholly sustained by the facts.

The Barred Rock people are making the most extreme claims for their favorites, not only as to fancy points, but as to laying capacity. I think the most extravagant claim of all is now made for a Barred Rock. That this variety has always suited farmers admirably is a known fact. It has usually given them winter eggs where they could not get them from other breeds. But has it given them enough? Just a few days previous to the present writing, I saw the claim made publicly that the Barred Rocks enjoy the distinction of being bred by a greater number of people and in greater numbers than any other variety of domestic fowls. Admitting that this statement is true in a somewhat modified sense, it points in a direction toward which, it seems to me, every Rock breeder has refrained from looking. *If* it be true that more persons breed Barred Rocks than breed any other varieties of hens, upon the

Barred Rock must fall most heavily the odium of having produced, in average hands, — so says the record — only six dozen eggs at the most, as an average. This statement cannot be gotten around. If most of our farm birds are Rocks, they are making a very poor showing for the breed as to laying capacity.

The White variety was a wonderful "find" for this country; this we may freely admit. I do not say that she is so much better than her Barred sister, except in the matter of dressing off for market better, when the pin-feathers are green. But she is a beauty, a show-room fashionable female, and she is a grand — to use a favorite term among breeders — a "grand, good bird." It is freely predicted, even among those whose opinions are not biassed, that she is the coming fowl, despite the tremendous and increasing competition of other varieties. And one who chooses White Rocks will make no serious error.

For the average worker, any buff variety is more difficult than a white sort. When the recent discussion was on before the poultry-law-giving body, and the question of comparative handicaps came up, each breeder of the plain colors (the ones handicapped when competing for points with the more difficult, party-colored varieties) held his own difficulties up for sympathetic public gaze. "We breeders of whites have brassiness and black flecking to fight"; "we breeders of blacks must struggle against white and against purple barring"; "we breeders of buffs must fight against black and white, and mealiness"; so their stories ran. Of course, they were all true: the question was only, which was the worst to fight. Every worker must find out for himself by trying or by testimony. Every variety has its difficult points.

The Buff Rock gives the same color troubles that all buff breeds insure. She is a valuable bird, and has a claim of having laid two hundred and thirteen eggs in her third-year form. The best layer we have ever had was a Buff Rock, a bird kept over from her first year because, and *only* because, she was a good mother. In her second year, she laid twice as well as during her first year. But she ran alone during much of her second laying summer. Any good hen under this condition, on range, is likely to develop "phenomenal" laying. On the whole, if either of the newer Rocks crowds aside the old variety, I think it will be the White Rock. She is a most excellent layer, and a market bird almost beyond criticism, according to the demands of American marketmen.

In dealing with birds of this type, one needs to keep in mind that they fatten easily, and that the fat is laid on internally as well as externally. Internal fat, if excessive, may crowd the organs of digestion and of reproduction, and a tendency to fat is quite likely to affect the liver as the fowl grows older. For this reason, the fattening ration that would not hurt a Leghorn, might break the Rock down utterly. The condiment that would just freely stimulate the Mediterranean type of bird, might bring on digestive difficulties, or inflammations anywhere along the egg duct. Any who may have had occasion to dissect animals will know how fat tends to make all the rest of the carcass tender. Any one who cooks, or any one who buys meats ought to see the bearing of this. Organs which are tender in texture must, of necessity, yield more easily to strain of any sort. Thus an extra large egg, which often goes with the tendency to fat, would the more easily rend the walls of the egg duct, and pass into the abdominal cavity, there to make fatal difficulty. This condition is so often found, and egglets cooked by inflammation are of such frequent occurrence in hens, that it behooves us to think whither these things point.

There is not much need to be afraid of fat in a young hen, especially if she be laying freely. But, given a breed that takes on fat with unusual ease, it is easy for the just-fat-enough condition that makes for a good egg product to pass into the too-fat condition, which makes for disease and disruption, especially after the first year.

The matter of good muscular development of the posterior parts is one of moment to the laying hen that is to do phenomenal or even noticeably good work. It is of more importance to any variety that has tendencies toward overdoing some of the good qualities. A study of these points is of vital importance to those, especially, who would carry the breeds that lay on fat very readily. A White Rock hen was shown at New York some time ago which tipped the scales at eleven pounds. She was bought by a young enthusiast to put into his breeding yard. I wonder if an old hand would have done it.

From one point of view, the Dominiques and Javas, good as they are, demand but little attention, because they have not, during a very long period, been able to make a wide public believe in their merits. Among all the American varieties and breeds, the Black Javas and the Dominiques claim precedence, and common report now has it that the beloved Plymouth Rock itself sprung from the union of Java and Dominique blood. Be that as it may, the Plymouth Rock has forged to the fore; the others have fallen well to the rear.

Personally I believe the Black Java to be as good an all-around bird as we have had, as far as performance while living is concerned; but a market fowl with dark legs cannot make headway against the dominating liking for yellow ones. I remember a farmer's wife, whom I knew in my youth, who searched during almost a lifetime for a black hen with yellow legs. I never heard that she found it. Her search had nothing to do with market demands, but expressed her own unbiassed preference. Professor George C. Watson, in his valuable book, "Farm Poultry," calls the Javas good winter layers, hardy, and true breeders; he says that they are easily raised, endure confinement well, and produce good broiler stock. Professor Watson is a most careful observer, and his word on any economic question pertaining to the farm is quite worth noting. The early Dominique was rather of the Leghorn type, the Java more like the Asiatics; the two were united in the hope of getting a better layer than either. The Dominiques are claimed by those who breed them to be steadily increasing in favor. A prominent breeder of the Dominiques asserts that they were "doubtless" bred long before the American Revolution. It is also claimed that they have been made especially hardy, through survival of the fittest in the olden days, and that they are now the hardiest and healthiest of all our varieties. But this claim has also been made for the Rocks and the Reds!

Age, hardiness, and all other good qualities, with a fair field through priority, have not made of the Dominique a leader. Though she is only a half pound lighter than the Rock, the latter has taken and kept the lead, and no one can with authority say why. It is as likely as not that the name "Plymouth Rock," with its suggestions of solidity and its appeal to the sentiment, has been quite a factor in its success. Again, the pushing power of a new thing well-backed is well known. No one who has bred the Dominique in later years has been known to acknowledge that it had any serious faults. It has recently been said that now a hundred "intelligent and appreciative breeders" are working to bring it to the fore. If this only is what it lacked, there may be a new story to tell, farther on.

THE WYANDOTTE

The Good Old Silvers — The Many Followers of the Silvers — The White Wyandotte and Her Claims — Competition Record of the Silver Variety — Golden, Buff, Partridge, Silver Pencilled, and Black Sorts — The Columbian — Color Problems

ENOUGH matter has been published in this country on the origin of the Wyandotte to fill volumes. "Facts"(?), theories, suppositions, go to make up the sum total of what has been said. Every little while the war of words breaks out anew, and the terms Sebright Cochin, Eureka, Excelsior, Silver Spangled Hamburg, and Chittagong are jumbled together till we who read are ready to tear our hair over the impossibility of making anything out of it any more sane than — "Katy did" — "She didn't." Summing it all up, we might as well settle down to the statement of the man who was Secretary of the American Poultry Association, both when they were refused admission to the Standard and when they were admitted. For they *were* refused at first as "not worthy," and the petitioners went back and worked on them till they made them worthy, as they have been working on them ever since to make them worthier.

Mr. J. Y. Bicknell, the Secretary mentioned, affirmed that no living man knew when they made their first appearance or knew anything definite as to their origin. "I know this from a careful and persistent search from every known source, when I was breeding them," were his words.

If we take up to-day the poultry books of the early eighties, we shall find cuts of a laced black-and-white bird, called "The Wyandotte." No longer is this "the" Wyandotte; that honor belongs to a white sport from the old, original Silver Laced, or, as she is now called, the "Silver Wyandotte."

But the Silver Wyandotte is not only a very valuable bird, she is also the progenitor of all the sports which bear the name of Wyandotte, and this is quite honor enough for any variety. Perhaps not all the varieties now listed as "Wyandottes" carry her blood in all their groups, but at least all have thought it worth while to adopt her name; also her shape — with some emphasis.

Of all her descendants, the most universally valuable, to date, has been the White Wyandotte. The sensational birds exhibited for some time past bear little resemblance in type to the Silver Wyandotte of the eighties, the head being really the best point of identification. The winner of first and special for shape in the cock class at the Boston show in 1907, and also winner of a bronze vase presented by the representative of the Japanese government as *the best bird in the show*, was a bird bearing virtually no straight line in his entire contour, except those which delineate his stout legs and toes. From head to shoulder, from back to sickles, down-curling; from tail to legs, from shanks upward round the almost pouting breast, the eye follows a succession of curves, the body forming almost a fat crescent. This bird is considered one of the nearest to the ideal, and he differs notably in outline at the rear from some which have won high honors in recent years. So far was the craze for a short body carried that birds were shown which looked as though they had been flattened, or knocked upward, at the highest posterior point. Some, indeed, appeared to have had feathers plucked from below the tail (the tail itself being undeveloped), in order to emphasize the appearance of shortness of body.

It is beginning to be recognized that this was a craze. Leaders are calling for a halt, and warningly crying that the utility value of the variety as an egg producer will be ruined, if the body is not lengthened. However this may be, the White Wyandotte, in the person of one of her breeders, claims a present record of two hundred and seventy-two eggs within one year, and she has made during many years past a record in the yards of hundreds of breeders which has enabled them to enjoy life and its amenities far better than they could have done without her. The Silver Wyandotte, with perhaps not so high a record, has a more highly honorable one, in that it was won in competition with hundreds of other birds under the same conditions and in public test. She finished at the head of one of the great Australian competitions with an average of two hundred and four eggs for the six birds in the pen, while five other pens of the same variety made from one hundred and forty-eight to nearly two hundred as averages. This is the best kind of record work, because the competition was against the fowls of many breeders and in many varieties: Asiatic, Mediterranean, English, and American in breed types.

The Golden Wyandotte is a beauty-bird which has fine utility qualities as well. Probably the fact that her colors do not please as well as others accounts for her lower rank among the Wyandottes. Other varieties are newer, and have hardly had time to crowd the originals yet; but public opinion is pretty near a unit in believing that no party-colored bird that is difficult to breed can hold its own against the white or the buff varieties in any breed. The Partridge Wyandotte is, possibly, the nearest of any to a genuine fanciers' bird; but it is too difficult to breed to Standard for it ever to become highly popular, and the bias in favor of the birds that dress off nice and clean looking, will keep the Buffs and the Whites in the lead. I presume the Silver Pencilled are at the foot of the list as to popularity, although

THE WYANDOTTE

the Columbians were last admitted. No black bird can hold its own against the really widely popular sorts, so that the Blacks, though undoubtedly good birds, may be dismissed without discussion.

One of the editors of perhaps our strongest poultry magazine has recently asserted that, when it comes to the test of genuine popularity, every breed must stand or fall upon its value as a market, or table, fowl. I am not sure whether or not he meant this to include its egg-laying capacity, but I can hardly think such a statement would be made without this inclusion. The multiplicity of varieties has almost whelmed the less striking, even of so good a breed as the Wyandotte, and the *created* popularity of some breeds and varieties has caused them to overshadow at times varieties that are really better. I am somewhat of the opinion that the Buff Wyandotte has suffered to some extent in this fashion, and that we have not yet heard the last word about her and her rank. Mr. Felch, than whom no man knows breeds better, both as to their rank as fancy fowls and as to their economic capacity, has stated positively that the Buff Wyandotte is the best layer of all the Wyandottes. It hardly seems that this could be true of a bird with Cochin ancestry; but it has often proved true that the Cochin, in furnishing the hardy, resistant frame, has made a good foundation on which to build the higher egg-laying qualities of other breeds.

The Buff Wyandotte is a handsome bird, either as dressed for the show, or as "dressed" for market; she is faithful at the egg-basket as well. And I think there may come a time when the mad rush for the white market fowl shall be stayed a little, and the Buff Wyandotte may then take a more prominent place than she now holds. Though it is in one sense high credit, it is in another sense a discredit for any one breed to have so many varieties, and I hope we may never have another breed with so many varieties as has the Wyandotte.

Never, since I have been familiar with Standard-bred poultry have I seen any variety come to the front so rapidly and with so little destructive criticism as has been the case with the Columbian Wyandotte. The Brahma feathering which she carries has always been popular, but the leg-feathering has driven many a man away from the Brahmas, who would otherwise have loved them. Given then a bird of the favorite Wyandotte type, familiarized to the world at large by seven good varieties, and having the well-liked Light Brahma plumage, minus leg-feathering, it would, of course, be expected to meet popular liking.

The variety has been widely praised, even among those who carry widely different varieties. It is not yet well selected, but has made so good a record that prophecy runs among the older seers that the "Columbian" bids fair to take the leading place amid those

THE WYANDOTTE

birds notable for both beauty and utility. More unifying work has yet to be put upon this breed. Selection and breeding must better its type, its feathering, its egg. In the meantime it has already proved an exceptionally good winter layer, and the good winter layer is what every poultryman from Maine to California is seeking. If bred to a farm type, rather large and long, as compared with the present Standard requirements, it may eventually crowd even the Barred Rock hard in the farmer's affections. All these newer varieties still need time, of course, to bring them up to the mental standard we are idealizing for them.

One of the best forward steps the Standard makers have ever taken was the extremely simple one of introducing a brief descriptive paragraph before giving the Standard points of a breed, showing how distinctiveness is attained, by calling attention to the why and how of specific difficult points. In connection with the Wyandottes, it refers to the interesting color problems met in handling each variety: in the Whites, to secure the pure white, avoiding cream and brassy tinges; in the Blacks, to avoid purple barring; in the Silvers, how to get silvery hackles without the brass in the white; big, beautiful, oval centres without mossiness, and breast plumage free from the "frosted" edge outside the black lacing; in the Goldens, how to get the right shade of golden bay which takes the place of the white in the Silvers; in the Buffs, how to get the color even and of the same shade throughout, without touches of black or white; in the pencilled varieties, how to get richness of color in the base, with a clear pencilling laid upon it.

One of the helps which the new Standard will have is a set of color plates, the best that the wonderful advance in color work will allow. These will not be breed plates, but accessory plates to show the proper tones in feathers for the various colors, which it is virtually impossible to describe in words. The question of proper illustrations has harried the American Poultry Association for a decade, and the next issue of the Standard will show some really artistic breed reproductions, which it is hoped to have true to ideal and as true to life as possible. Camera and pencil will combine their resources to offer the engravers a proper base for their work.

The color question has always been a difficult one, especially for beginners. There is the getting it, not only, but the keeping it. The following letter shows some of the difficulties.

Though referring to Orpingtons, it applies as well to all buff and white varieties, whatever the breed: —

"I have been keeping Buff Orpingtons for two years; they lay pretty well and are grand table chickens, but I find the buff fades. Though I have a fine flock, if I have hens of different ages I have hens of different shades, and some fade

THE WYANDOTTE

sooner than others. In consequence, the flock looks uneven, and I think that detracts from its beauty. Do all buff birds fade? Is buff more popular than white? Are Buff Orpingtons brought to greater perfection than White or Black? I want to weed out my poor layers with trap nests, and I find that some of the best-colored birds are the poorest layers. I would like to try Whites. Have they any glaring defects? Do black feathers crop out? Are White Orpingtons pure white? It seems to me the great egg producers of the world that are also prize winners are white. Do you think I can improve on white for a combined utility and show bird? I want to know, if I change, what I shall be up against."

Reply: A good many years ago Mr. Felch stated, in one of his books, the general law that all breeds grow lighter with age. This referred not only to the fading of the plumage annually, as it nears the period of casting off, but also to the lightening of the color with each successive moult. So you will see that you have not much to complain of in the simple fact that your Buffs fade. I think it is true that buff fades more than many of the other colors, and it is also true that buff birds vary much, not only in the color of individuals as compared with other individuals, but in the feathers of a single bird as compared with other feathers on the same bird. Whenever there is moulting, there is, in an especial degree, unevenness of color, because new feathers are strong in color and older ones are more or less faded. That is, unevenness of color is very largely due to difference in the age of the feathers. Even where the birds are not faded, it is only the exceptionally good flocks that contain birds largely duplicates in color as some come light and others medium or dark, at their first complete feathering.

All these facts together make the color question with Buffs a serious one, even though they are solid colored. And when you put in the requirement that your buff birds shall be first-class for both utility and show, you are putting a snag in your own path; since the standard way to control the color of buffs is to keep them out of the sun as much as possible, and this cannot make for the vigor demanded in the best utility birds. There will always be some points on which "utility" and "fancy" will quarrel so seriously that there seems to be no reconciling them. We must give up one or the other, or else compromise. But I am certain that the longer any one breeds fancy stock, the more averse he will be to subordinate vigor, even to some greatly desired fancy point. Thus it comes about that many breeders of white birds stop at the point where the fancy requires them to produce birds that are chalky, in a breed having yellow legs and skin. Such a breeder can hardly reach the top as a breeder of exhibition birds, but he will have better success along other lines, and may have a better conscience; because a goodly proportion of any breeder's customers are of those who cannot hope to have any success with birds of less than the fullest vigor, and any one who has bred delicate fancy stock long or practised double mating long knows that his customers are often throwing away money when they buy of him.

The question of brassiness and that of troublesome black feathers takes the place of the problem of fading and mealiness in buffs, as soon as one changes from the buffs to the white breeds. So you will do well to think pretty carefully before you change because of just one count against the breed you now carry. I do not say this because you are thinking of Orpingtons, or with reference to any special breed, but as a bit of common sense generally applicable. However, if you changed from Buff Orpingtons to White Orpingtons, which I gather is what you have in mind, you would probably meet but little trouble, because the latter are not required to have the yellow skin and legs common to the American white birds.

I don't think I quite get your point of view where you state a belief that the great egg producers of the world that are also prize winners are of the white varieties. I see nothing to lead one to think that prolificacy, prize winning, and white feathering are necessarily linked. Perhaps the fact that white birds are in the large majority, being much more popular than any others, gives you the idea. And when three of the tremendously popular breeds in this country, viz.,

Wyandottes, Leghorns, and Rocks, are so very widely distributed in their white varieties, one might easily draw conclusions that have no basis except in the fact that such a large number of white birds is raised. It is, I think, a world axiom that white in itself indicates a lessening of vigor, however slight, as compared with darker colors in the same lines of blood.

However, I do not think you "can improve on white for a combined utility and show bird," taking this question on its sole merits. Aside from the fact that white birds show the soiling of the plumage so quickly and so much, there are few arguments strongly against them. You would, of course, be "up against" the hot competition of the whole country, and you could avoid this entirely by choosing, as a few do, some breed that is little raised. As headquarters for some such variety, one stands some show of custom, and plenty of "show" in the show-room. But there is nothing like so much interest in the whole thing, and perhaps the number of customers would not be equal to your share of those who are buying the more popular varieties. The popularity of white birds is inherent in the make-up of the people; the "popularity" of buffs, on the other hand, is a worked up popularity, such as it is. It arises from a supposition that buffs are going to be first favorites in some not too far distant future, based on a belief that they are both very much liked and fashionable now. This is true, but true in a limited sense; and many of those who take up buffs do so more because of prophecy than of present fact. Perhaps I don't need to tell you that prophecy has far more of the characteristics of an art than of a science. It is freely prophesied that the Buff Orpington is the coming bird. It is also prophesied that the White Rock will at no very distant date be the most popular variety of any breed in this country. But prophets have their limitations of vision. Some have axes to grind; some think themselves especially favored with quick vision of the future when they are not; some speak without full knowledge of the facts, etc. It is a great thing to be able to look ahead, but we all are mistaken sometimes.

THE RHODE ISLAND RED AND THE BUCKEYE

"The Ideal Breed"? — Origin — General Style — The Pea-combed Red Birds — The Buckeye a "Made" Breed — Beauty of Coloring

THERE is one Rhode Island Red breeder who has had my special interest because he has succeeded, by his own account, in doing what no one else has been able to do with any other breed known! I mention this to show how far enthusiasm may lead a fancier. This is his statement: "When my birds are grown I examine each one carefully, and if it fails to coincide with the Standard in any way, I throw it out of the pen and send it to market." No other breeder, no editor, no breezy amateur with his frequent gush, no judge in his linen ulster, has ever found a bird of any breed that coincided exactly with the Standard description. This man stands alone. And on the strength of this he sells all his eggs — so he says — to his neighbors at two dollars a sitting and has no difficulty in thus disposing of every egg! Please notice that this is the breeder's own exact statement. In his

THE RHODE ISLAND RED AND BUCKEYE 37

enthusiasm for his favorite breed I do not suppose he realized that he was putting a direct untruth into his "article on Rhode Island Reds." Yet every other breeder would know that it was gross exaggeration — an impossibility.

In another article in the same publication concerning this "ideal breed" occurs this statement: "As there has been so much outbreeding done with the Reds, and consequently they have acquired so much foreign blood, I am willing, as yet, to sacrifice a little in comb, eye, or shape, in order to obtain something fine in feather, because I have found the Reds harder to breed for feather than for other Standard characteristics." Here is a man who is telling the straight truth, and no one who is deep in fancy poultry mating and breeding will think the worse of the breed for what he says, because the same thing is more or less true of virtually all breeds. The Red is a little worse because it has so many breeds behind it in its blood.

But now put this statement beside one which appeared with a query a few days ago, to the effect that Rhode Island Red eggs at ten dollars a hundred from a well-known breeder gave birds "of three or four colors, and a great many of them have side sprigs" (on the comb). It is easy to see that the outcrossing mentioned by the breeder above accounts at once for the variations in color complained of by the novice buyer. And there are few, if any, breeds of this difficult color of which the same is not likely to be measurably true; yet, in the face of this fact, it is also true that birds of such evenness and depth and richness of color are selected for the big shows that expert and novice alike simply rave over them. The finest ones are indeed of rare beauty, and the adjective is all the more fitting in that it also expresses the fact that such birds are also "rare."

Theoretically, it is a beautiful idea — this of an ideal bird! I do not know that any have yet claimed to have found it, except fanciers with birds to sell; but it is the bird we are all looking to find some day, in hope, at least. This breed would suit us, with an "if." That one would be perfect, with two or three "ifs," etc. Still, I do not suppose that any breed now exists that has not been claimed by some one to be "the ideal breed." There are even now popular breeds, unpopular breeds, and breeds of which ninety-nine persons out of one hundred never heard the name, to all of which some enthusiast has at some time applied the desirable adjective "ideal."

If you wish to go back to the beginning, when the Rhode Island Reds were called "the Bill Tripp" fowls (after the man who originated them, and had them at their then best), you will find yourself face to face with Asiatics, while our red breeds are known as American breeds; but even here, though the story of their origin comes from the children of the originator, there seems to be a little doubt or difference. The son of William Tripp has said that the foundation of the breed was the old Chittagongs and Cochin-Chinas, crossed together. The daughter states that she remembers the incident of her father bringing the original red rooster home; and Dr. Aldrich, a prominent Rhode Island Red breeder, who went to her for information, said of the fowls, "They were undoubtedly of the old red Malay and Shanghai blood." But William Tripp crossed and recrossed among the descendants of the old red rooster till even he, probably, would have found it difficult or impossible to tell the exact blood entering into the more modern specimens of his own breeding; not to mention what they went through in other hands than his. Until the last years of his life he did not breed for points or ancestry, but rather for featherless yellow legs and a good carcass for market, with superior laying quality. The later crosses, in the hands of neighboring farmers, and fanciers' efforts at improvement after the birds had a definite breed name, have rendered the breed a remarkably varying one as to different strains and have involved it in obscurity not now to be cleared. We must be content to accept them for what they are and what they may become from this time forth.

Although so widely bred, it was not until recently that the Single Comb Rhode Island Reds gained admission to The Standard of Perfection. They are the Rhode Island Reds concerning which, in an explanatory note preceding the requirements, The Standard of Perfection says: "They are believed to have originated from crosses of the Asiatics, Mediterraneans, and Games." It hardly needs to be said that a breed made up of three widely differing types, as well as of the blood of many different breeds with different mix-ups in different hands, will need many years to overcome variation of type, if, indeed, this is ever overcome. For a breed that has been in existence fifty years (as is sometimes claimed), the Rhode Island Reds will produce an incredible number of rank culls, as well as some of the most beautiful birds that now appear at our exhibitions.

The Standard calls for a bird of good size, running from five pounds in the pullets to eight and one-half pounds in the matured males. A body broad and deep and long must necessarily, it would seem, weigh even more than the extreme of eight and one-half pounds for which The Standard of Perfection calls; and when the demand is also for a long keel, extending well forward, so as to present an oblong appearance, and also for a tail carried at but forty degrees above horizontal for the sake of adding apparent length, there is

THE RHODE ISLAND RED AND BUCKEYE 41

the foundation not only of a big bird but of one that looks larger than it really is. I am quite curious to see what will result in ten years more from these requirements. The tail of the female is to be carried five degrees lower still than that of the male, as noted above. Thighs, shanks, toes, and tail in the male are to be of medium length, while the tail in the female is to be rather short. It seems a question to me whether a bird with not only all its body measurements but its contour in excess (for the breast is to be full) and all its supports and outportions medium or short would not be a good deal of a monstrosity, if bred strictly to the letter of its standard. Most other requirements, aside from those mentioned, are for medium development.

The comb must be straight, not lopping, and must have only five nice, even serrations. The almond shape is called for in ear lobes, and they are rather small and fine. Nearly all requirements in the various breeds are based on a mental standard, only to be gained through experience with Standard bred fowls. For instance, a back of medium length means a back that is medium as the backs of fowls average. So in most other requirements, one must have an idea in mind of what an average bird in each of these points is. This average is the mental standard of comparison.

The red color is kept a prominent characteristic of

the breed throughout. Eyes, face, wattles, lobes, under-color, are to be red. Even the beak is to be horn color, with a reddish cast. Shanks and toes, reddish horn, when not yellow, and it is considered desirable to have a line of red showing down the sides of the legs. Lucky Rhode Island breeders! For most breeds have to get rid of this line, which is continually trying to appear in many of them.

As to plumage, the red is to be rich and brilliant, and as even as possible, darker color on wings and back being allowable, but mealiness and shafting being prohibited. The color should be so brilliant as to look glossy. Under-color is to be red or salmon, with no smut, and if several specimens are equal in all else, the prize will go to the bird best in under-color. Webster defines "salmon" as "reddish yellow, or orange color, like the flesh of the salmon." The flesh of the salmon varies a good many shades, and the redder the "salmon" the better, probably, it will please the judges.

Any one may see that, as good birds become more common, one who would breed show birds must be more and more sure to know every little point of a good bird, as upon these little points often depends the capture of the prize. Even "laundrying" the bird, as it is sometimes called, somewhat in derision, often makes the difference which gains the prize. At one

THE RHODE ISLAND RED AND BUCKEYE 43

Boston show carelessness of the transportation folk ruined the birds of a prominent breeder, to the extent of losing him some coveted firsts, and as small a thing throws some one out of a prize nearly every time a big show occurs.

The admission of two red varieties of American origin to the American Standard of Perfection at a single meeting gave us enough recognized red breeds to form a class by themselves. And as the color is rather likely to prove more generally popular than any other except white, the red breeds promise to be unusually interesting for some time to come.

The "American Reds," having been well known to the public for a considerable time under the name of "Rose Comb Rhode Island Reds," performed a grand transformation act not long before the Minneapolis meeting of the American Poultry Association, and asked for admission under a name by which they had not before been recognized. They had been shown many times as Rose Comb Rhode Island Reds.

When the report of the meeting went out, the breeders of the original variety were up in arms. Doubtless they would have preferred not to have Rose Combs of their breed. The Rose Combs had pushed ahead wonderfully, and crowded the Single Comb Rhode Island Reds (themselves admitted but a few months earlier) rather hard. (At some of the largest shows, the Rose Combs have occasionally been more numerous than the Single Combs.) When these same Rose Combs asked to be admitted in a way that made them keener rivals, without acknowledging their debt to the Rhode Island Red breed, the matter seemed to have reached the status of direct injury.

I am mentioning this only to forefend puzzled questions in the future from some who may stumble upon mention of "The American Reds" and wonder how to place them. It will be seen that these are the same as the Rose Comb Rhode Island Reds. They were not able to hold the name against the storm of protest. They occupied the centre of the field for one year. During that time, Eastern breeders who had not taken the trouble to go so far as Minneapolis to attend the American Poultry Association meeting, made matters very lively. They raised such a protest and clamor for reversal that the noise of it was heard across the continents. It was kept up for a year, and at the following annual meeting of the American Poultry Association everybody begged everybody's pardon and the name "American Reds" was dropped. The name was carried in advertisements for a short time, but has now passed out of sight.

Buckeyes are a different story. There is quite a long tale of a mix-up at the beginning which caused the Buckeyes to be confused in the minds of the public

THE RHODE ISLAND RED AND BUCKEYE 45

with the Rhode Island Reds. They are, however, of totally different origin, and do not carry the same blood, except as breeders have tried to make a Pea Comb Rhode Island Red by using the Buckeye blood. The Buckeyes were made from the union of the Plymouth Rock, the Buff Cochin, the Game, and the Cornish Indian blood.

It is a somewhat curious commentary on circumstances that, while Rhode Island Red breeders have dallied and opposed and refused recognition to the Pea Comb type of this fowl, a worker in the Middle West should have been quietly and persistently toiling and planning to produce a bird which, when first brought to public notice, could almost have been mistaken for the Rhode Island Red, but which had a pea comb. Such men as Mr. Felch have stood for the assertion that the Pea Comb was the only consistent type for Rhode Island Red breeders to follow, and the mahogany red the true and desirable color. Yet the breeders turned away from this expert advice and went their own ways.

The name Buckeye Reds, as first given, was partly for the Buckeye state, where they were originated, and partly for the color of the females, which is almost that of the ripe buckeye, or, as easterners know it, a horse-chestnut. The gloss and color of the buckeye are charming, and, when transferred to moving, active birds, with the contrasting red blossoms of combs, cannot fail to be attractive. The originator of this new breed is a woman, who describes herself as "of the greyhound order" — that keen, slim, nervous type that "brings things to pass." She counselled with two of the best judges of the Middle West, one being secretary of the American Poultry Association, both of whom advised her to go ahead, get her breed admitted to the Standard as soon as possible, and give it to the country. The favorite Plymouth Rock blood enters into the new breed, being the first base used for crossing. There were two infusions of Game blood, which always gives good utility quality in a cross. The Standard weights run from five pounds in the pullet to nine pounds in the matured male. The male body plumage is described as "dark, rich, velvety red, garnet, or dark cardinal in shade." The tail may have darker points. Legs and skin are of the desired rich yellow. "The skin is not tough and leathery, like that of the Rocks, while the flesh is finer grained and of better flavor." This is probably a heritage from the Game ancestry.

The originator says: "I thought the Langshans the handsomest birds on earth. They laid well, too, and the white-skinned flesh was tender and delicious. The black pin-feathers, however, nearly drove us to distraction, so we switched off to Barred Rocks, which we

THE RHODE ISLAND RED AND BUCKEYE

liked very much, although we never made great layers of them. Thinking to improve them, we crossed them with Buff Cochins, then used what we supposed were Black-breasted Red Games with them. As part of these Games had yellow legs and pea combs, we now feel sure that they had been mixed with the Cornish Indian Game before we got them, and here is where we got the pea comb.

"This mix-up produced a bird or two that were red as foxes, with yellow legs, and I went wild over them and conceived the idea of raising a whole flock like them. When at last I had a sizable flock, a name was the next puzzle. We finally condescended to give the honor to our own State and call them 'Buckeye Reds.' This seemed the more suitable as the hens were as brown as a ripe buckeye, with a red sheen in the sun."

When the breed is perfected, the birds will be darker than the Rhode Island Reds are now. The pullets will be nearly a mahogany red, while the males will be more brilliant. They have had much of the Cornish style, but the Standard calls for medium legs, neck, etc., so that selection will soon weed this out of them. Indeed, a fair proportion do come with as short legs as one would care to see on them. They are good fighters, but singularly chummy and confidential with people, being easily tamed. I do not see how this can be the case, but it is entirely true. Those we had made the most desirable sitters we have seen. They have a small pea comb, and I kept them two winters in what is practically an open shed, without seeing a sign of frozen comb or wattles. The last winter they were free to go out and in at will nearly every day. When we first got them, and for some months, I felt undecided as to their superior value. The word that fits them is one not usually considered very strong, yet it can be applied only to such varieties of fowl, beast, and plant as please many people in many characteristics. That word is "satisfactory." The rose catalogued as "satisfactory" is the one most people will finally "tie to"; so with the satisfactory animal or fowl; it is better for the ordinary worker than one which has one or a few phenomenally good qualities and a lot of poor ones.

A Buckeye cockerel killed for the table, after being bled, tipped the scales at seven pounds, and measured four inches straight across the breast. A breast like that makes good slicing, and such a bird looks like a young turkey when it comes on the table. The meat is of first quality.

In several respects, the Buckeye standard reads much like that of the Rhode Island Reds. The plumage, however, is described as a red of dark, rich, and velvety character, even running to garnet or dark cardinal, almost mahogany red. The under-color is to be of a lighter red, with some slate in the back. Black is allowed in surface color in tail and in under folded primaries. Other things being about on a par, the preference is given to the bird having the most even dark red surface color. A good judge has said that the Buckeye standard is much the best standard among the red breeds.

I think the majority of people fail to realize that more allowance must be made for a new breed than for an old one — that is, one cannot be sure of so large a proportion of good birds from the number of eggs — I mean good from the Standard point of view. Selection must be rigid, or bad faults crop out. Some have tried to throw public discredit on the Buckeyes because a good many feathered legs, off-colored feathers, etc., appear, but the same could be said, even to this day, about the White Wyandottes, the Leghorns (old as they are), etc., and eminently so of the Rhode Island Reds. The trouble is, breeders who want to try experiments are continually introducing foreign blood, which in time goes all over the country, and no flock into which outside blood is brought can be safe from such contamination. Some time ago I had Rhode Island Red eggs from a breeder of considerable pretension, who gets a fair proportion of first prizes and advertises that his stock is unequalled, or words to that effect. Ever so many of them had feathers on the legs when hatched —

more, I think, in proportion, than any Buckeyes I have yet seen.

Twice I have had communications saying that the writer thereof had been "beat" in buying Buckeye eggs. It is easy to put out fraudulent goods, especially in a new breed. But it is possible that a buyer might think himself "beat" simply because he made no allowances; or when he had bought second or third grade eggs. Then, at times, in any breed, a male which is a model specimen will not give good progeny. One of the best males, in appearance, which the originator of the Buckeyes ever had produced chicks with one particular serious fault. Another, a much admired male, produced only disqualified chicks on the male side. All these things have to be taken into account before one cries "fraud."

On the general farm, conditions are such that fowls tend constantly to become smaller in size. For this reason the good-sized Buckeye, with a male weighing nine pounds when mature, should make an excellent farm fowl.

The best specimens of the red breeds are now so very rich in color that no one could help admiring them. The under-color, quills, skin, etc., are so nearly alike that they dress off beautifully. There is no shadow of doubt that American breeds suit Americans full well. Nor is there much doubt that the handsome red breeds,

THE RHODE ISLAND RED AND BUCKEYE

being solid in color, will suit better, when well known, than any party-colored breed. The Buckeyes that I have seen, from the originator's yards seem decidedly nearer the Malay type than any of the other reds. This shows partly in the greater difference in size between the males and females.

THE CHIEF COMPETITORS OF THE AMERICAN HEN

The Leghorn's Place — Orpington History — Showroom Popularity — The Orpington and the Wyandotte

EVERY one who knows even a little about poultry knows that the high standing of the Leghorn fowl cannot be controverted. This breed has made, and holds with tenacity, a royal place in the regard of the poultry world. Nevertheless, it has its detractors, its despisers, and its legions of grumbling friends, who admit its value, but decry the fact that it is not a general-purpose bird. This, in one sense, narrows the real rivals of the American fowls to one variety. For, whatever the imminent future may have to say about the "Crystal Whites," etc., to-day, the Buff Orpington is the only one that can rightfully claim the position of a notable rival of our own American-made birds.

Despite the fact that its first bid for popular favor as a distinct variety was made on the reputation of its predecessor, the Black Orpington, and that it is allowed to be an English-made bird, the history of the Buff Orpington is, in many senses, the history of a scrub. The originator of the Black Orpington claimed to have produced the Buff Orpington, as well, at his place in

Kent, England. Edward Brown, prominent in England as an authority on poultry, and lecturer on aviculture at Reading University, refers to the Kent production as an "assumed" new breed, and says: "Even if we accept the statement that some of the Buff Orpingtons were produced in Kent, there is abundance of evidence that the great majority of present-day Buffs are directly bred from 'Lincolnshire Buffs,' without the slightest relation to Mr. Cook's strain." Like our own Rhode Island Red, the "Lincolnshire Buff" was a specially good type of farmers' fowl, but in a very crude, shifting state, and not deserving the name of a breed, at the time the Buff Orpington was produced. Mr. Brown calls the Buff Orpington "a refined Lincolnshire Buff," and cites Mr. Cook's own admission to show that it had none of the blood of the Black Orpington, on whose breed name it traded. But it is a fact well known to breeders that, a few years after the introduction of any variety, it becomes impossible for any man to say with authority just what elements of blood enter into its make-up. Every man who breeds it may juggle with it, and we know that many fanciers do this continually with all their varieties, in the effort to gain some desired point. In this country, an English breed soon loses some of its former characteristics, in adapting itself to the new environment. The first exploiter of this particular variety has himself bred and pushed it in this country, — a very unusual happening to any variety. We may therefore assume that our Buff Orpingtons are measurably of English type still, except in case the originator has himself modified them to make them stronger competitors of the true American type. I do not say that this has been done, but it is a thing that would be likely to occur. Relative to this, an American woman once told me that she bearded the originator after this fashion: He was giving her a personally conducted view of some of his birds, on exhibit in New York's largest show, and evidently aiming to make a strong impression. "Why, Mr. C., I have *cooked many a better bird than these*," was the unabashed comment of the American poultry breeder — an expert usually in what she takes up. This might signify only that her ideal of the White Orpington (the bird in question) was different from his. But at least she believed herself to be speaking to the question of the American ideal of the White Orpington.

Mr. Brown says that no variety has suffered more from the common tendency of breeders to exaggerate the characteristics and qualities of their favorites, than has the Orpington; clinching the statement by adding: "If one tithe of what has been said respecting them were true, they would deserve to be canonized." This, it is to be remembered, is the authoritative word from the home of the Orpingtons, England itself.

Nevertheless, it appears that the Buff Orpingtons are making good in this country, their one notable failure to fill the American demand for a general-purpose fowl having to do with the white legs. They are a half pound heavier, right through, than the Rocks, cocks going to ten pounds. In all the essential points, the Standard descriptions of the two breeds are virtually interchangeable, the qualifying words "rather" and "moderately" in the Rock description permitting the greater length and plumpness needed to get the extra half pound on the Orpington. I cannot see where the Orpington could have the ghost of a chance to prove a victorious rival over our own buff and red breeds, unless it could establish conclusively the fact of being a better layer. This it has not yet done, and it would be a most difficult venture.

The show records as to numbers are an index of one kind of popularity, and in 1909 the Orpingtons as a breed stood well toward the top, with forty-eight hundred and fifty-nine specimens exhibited in the United States and Canada during the year. The three favorite American breeds and the Leghorns alone outclassed them. As far back as 1906, the Buffs alone, at Madison Square Garden, presented nearly two hundred specimens for the judges' consideration. The popularity of the Orpingtons of the Buff variety in England has been attributed to their possession of the rare combination, white legs and skin, with the characteristic of laying tinted eggs. The Whites are described as in all respects the same as the Buffs, except in plumage and beak.

In this country I think there is little doubt that the popularity of the Buff Orpington depends not so much on its professed superiority as upon the fact that, being a general-purpose fowl of the type generally suited to American notions of a profitable fowl, it has also been most persistently pushed. It is also a handsome, massive show bird, which fact always makes one kind of popularity for a variety.

Such testimony as to the comparative egg production as is available and counted fully reliable tends overwhelmingly to place the Black Orpington in the leading place, as far as the varieties of this breed are concerned. In one of the great competitions in the southern hemisphere, Black Orpingtons were backed by their owners as ten to one, over the Buff Orpingtons, as layers. That is, nearly twenty pens of Blacks were entered, and but two of Buffs. The birds upheld their owners' estimate very fairly indeed, the one hundred and twenty Black Orpingtons standing third for the breed prize and averaging above one hundred and seventy-eight eggs for the year. The Buffs, under the same treatment, averaged but one hundred and fifty-nine, and while this is quite sufficient to put them into the ranks of good layers, it does not advertise them favorably as against the one hundred and seventy-eight average of the Blacks. There were three entries of Langshan pens, which made a breed average of one hundred and eighty-eight. A pen of Langshans won first, and brought in the most money, but another pen did so poorly as to pull down the breed average to second place. One of the things averred to have been shown by this series of tests was that the newer breeds have not sufficient stamina to compete with the older ones. I should call this a pointer rather than a proof, but it may be worth thinking upon for a bit.

America — and it may be said the whole world — is looking for a variety that can outlay our present birds, and do it without loss in numbers of the best layers. It is of doubtful value to get a variety that will outlay existing sorts, if at the same time the mortality is heavy. If we cannot get this needed stamina in a new sort, what then? Our friends across the sea are tentatively saying to us that we are more likely to find the three-hundred-egg hen by selection from older breeds than by the making of new ones and claiming them to be superior. But the fact that a large entry of Black Orpingtons won third place on breed *average* somewhat tends, in my estimation, to dispute this conclusion, since the Orpingtons, though introduced some time ago, still rank, by comparison with most others, as a "new" breed.

Since writing the above, I am in receipt of tables which show statistics of Orpington exhibits in our larger shows for all the years since and including 1901. These make the entire number of Orpingtons shown (including six varieties and an "any other variety" class) to be well toward four thousand birds. New York alone reported nearly twenty-five hundred. New York, in 1909, alone reported four hundred and ninety-six, which is getting well toward the numbers at the Crystal Palace show itself, in England. In 1907, for some reason not known to me, the exhibit of Buffs at New York fell far below normal, while the exhibit of Blacks was higher than in any year before or since. In all other years the Buffs led, being usually half more than the Blacks and two or three times as many as the Whites. In January, 1909, there were two hundred and twenty-three Buffs shown at New York. But the numbers shown at Chicago and Boston bear no comparison with those shown at New York. These tables lack a part of the Chicago records, but show, in all, seventeen hundred and thirty-eight Buff Orpingtons in the three big shows for the period under consideration.

If it is possible for extra-good advertising to push a variety to which we oppose a pure American variety of almost the same color to the point of thousands of birds on public exhibition in three shows, during less than a decade, we need to look sharply to our *advertising*. If we are, instead, to look for this advance to the *quality* of the variety, we need to look to the *quality* of our own buff breeds.

I am rather sure that nine out of ten who read these words will not have known, before that reading, that an American breeder and judge of more than fifty years' experience "on the inside" of our poultry history makes the positive statement that, of all our Wyandottes, the Buffs not only lay the greatest number of eggs, but the largest eggs. "They lay more and larger eggs than do any other varieties that march under the Wyandotte banner." He also intimates that there is much Rock blood in many varieties of Wyandottes, in some breeders' hands, in the statement: "The Pencilled, the Partridge, the Buff, the White Rocks, and the Black Javas that come rose-combed are called Wyandottes."

Since we have such an American variety, in the favorite buff color and the favorite American type, to oppose to this competitor, which has proven itself either very strong or in strong hands (speaking commercially), we certainly ought not to fear such competition, provided that we make full use of our lead and our opportunity.

THE FIVE-DOLLAR-A-YEAR HEN

The Rose Comb Brown Leghorn made Her Record for this Breed-appellation — Her Merit — Her Popularity — Her Importance

AT the present writing, the poultry periodicals generally are carrying a modern advertisement claiming that it is easily possible to make a large flock of hens average more than seven dollars a year, and that it has been thus easily done. But the fact remains that a widely-heralded test, with the best breeds of the world in competition, was unable to reach this distinction. The eyes of the world were upon the affair, and all closely-interested men ready to herald to the world such success as might be achieved, yet $5.27 was the highest record profit, at market prices. This actual record breed-laying was done by the Rose Comb Brown Leghorns (only a few of these being in competition), and the record made was $5.27 *above cost of feed*. It was stated at that time that the actual cost of feed for this breed was just about half the average cost per pen. At the end of the fifth competition, with hundreds of birds of all breeds in the test, the *average* profit for all the hens had not passed, in any one year, above $2.91. This report of averages was first published in this country, I believe, in July, 1907. The year that the Rose Comb Brown Leghorns made this breed record, the Silver Wyandottes made their high average pen record of two hundred and eighteen. Our own Bureau of Animal Industry in its report noted those records, saying: "The winner, from an economic point of view, however, was a pen of Brown Leghorns from America. They averaged over two hundred eggs per hen, and as profit-givers stood alone, as they produced their great tally of eggs on half the average quantity of food consumed."

There is no doubt that the Leghorn fowls stand even higher in public estimation as egg layers at the present time than has ever been the case at any time previous. In the recent two-year Australian competition, White Leghorns took a decided lead. It has been said that fowls are more prolific in England, and probably also in Australia, than here. And Mr. Purvis, who spent a good many months with his finger on the pulse of poultry raising on the Pacific coast while residing there temporarily, has predicted that if this record is ever equalled in this country, it will be by Pacific coast hens. It is worthy of note that during the one year of the series of Australian tests that was distinctly unfavorable as to weather, the Silver Wyandottes won. Professor

Thompson stated his belief that in unfavorable seasons the Wyandottes and Langshans would be found equal to the Leghorns, as a general proposition. In his opinion, his tests narrowed the list of "best" breeds down to Leghorns, Wyandottes, Orpingtons, and Langshans, with not all the *varieties* in these breeds favored. "Laying strains should not be perpetuated, and bad strains for ovarian weakness should not be bred from," is his dictum. The first half of it is pretty sweeping, and the fight is likely to rage long around this point. There is one man in this country who has reduced his convictions on this head to a working formula. He is now manager of a plant carrying a good many thousands of hens, and he states his position thus: "It is an error to breed from an 'abnormal' hen. By 'abnormal,' I mean a hen whose laying record is far removed from the average ('normal') of the flock of which she is a member. When our flock average for last year was one hundred and sixty-two eggs per hen, we considered all hens varying more than ten per cent from the record 'abnormal.' This applied to those laying above, as well as those laying below the average, and we considered hens laying uniform eggs between the numbers one hundred and seventy and one hundred and eighty the best breeders in the flock." Mr. Ellis's experiment is on a large scale, and primarily for business development, rather than to affect an open-mouthed public. Nevertheless, the public will sit up and take notice of such results as he may report.

I have been breeding fancy stock, in a good many breeds, for a period of between twenty and thirty years, and I have not yet seen a breed that has been able to push the Rose Comb Brown Leghorn out of my yards. Grumblers will grumble that she is not large enough. This point is well taken, if heavy roasters are wanted, for the females seldom go above four and one-half pounds, and the average of good hens will probably be near four pounds. But the carcass is so plump, at every age, and the flesh so well distributed on this round body, that this variety makes the very best bird I know of for the small family dinner. The quality is of the very best, and, in our own family, this is always the favorite kind for table use. If it is a Rose Comb Brown Leghorn, there is never a question but that it will be good. I sell the birds to private custom with more confidence than is the case with any other variety. And whatever be the case with other trade, in our own private custom the call is nearly always for a bird not far from four pounds in weight. Roasters would probably run a little heavier, but we have much more call for "fricassee chickens" than for roasters. And this call the Rose Comb Brown Leghorn can meet most admirably. We never carry either the general-purpose fowl or the Leghorn alone, as each supplements the other to some extent.

The "five-dollar-a-year fowl," however, does not make her real fight for public favor on her own market value, but on the number and size of the eggs which she produces. These have averaged, in confinement, larger in size and greater in numbers than those of most of the other breeds we have handled. It is possible that the Columbian Wyandotte will push her to the wall, but she has not yet been able to do it. The little "Brownie" lays an egg that averages considerably larger than that of the Columbian, though the latter should be, on the average, more than two pounds heavier in body. It is considered a white egg, though it does not show the chalky whiteness of the eggs of some of the White Leghorns. And it is produced in such numbers that most buyers take the trouble to write and express their enthusiasm as to the value of the Rose Comb Brown Leghorn.

Although now shown in very goodly numbers, the Rose Comb Brown Leghorn makes very little ripple as a fancy fowl. It would not be necessary for her to enter this field at all, but that the variety clubs and the show lists form a sort of publicity bureau for all breeds, that can ill be spared. It is of little moment to the world at large that any variety should have so many points of merit as this one, *if* the world knows nothing of the facts. There are no breeders more enthusiastic than those who hold to this variety, but their enthusiasm comes primarily from a knowledge of her performances, rather than a belief that all the world should give the "Brownie" the palm as a fancy fowl. Yet the coloring, in fine specimens, is always noticed by visitors with words of much admiration, and the finer pencilling and other points are difficult to get. Despite this, the average worth of all specimens bred is really high, so that one who is not breeding to exhibition form need throw out very few specimens, if he have a good strain. Like other Leghorns, they are not so well suited for early winter laying as are the American breeds, notably the Columbian Wyandotte.

The chicks are admitted by all who see them to be "the dearest ever," and they are almost exactly like little partridges. Though appearing much like the Partridge Wyandotte chicks, except as to form, they are much prettier about the head. This is as it should be, since the head points of a show Leghorn have large value, in proportion to the rest of the body.

From no other buyers have I had so many enthusiastic letters as from the buyers of Rose Comb Brown Leghorns. I might probably except the buyers of Indian Runner ducks, but the Runners are so new to this country that the comparison can hardly be made. There is always something peculiarly attractive in a pullet just at laying maturity, but no other pullets have been so cherished here as the graceful, sprightly Leghorn pullets. And, although the Leghorn has the reputation of being very nervous and wild, she becomes most friendly and confidential with a calm-natured, quiet owner.

The five-dollar-a-year hen becomes of greater importance with the advent of higher prices for feed, than she has ever been before. For, while it is always joyful to take in a goodly income, it is a necessity that it shall represent a paying margin above cost of feed. The hundreds of hens in the earlier Australian tests — notably the first, when they averaged but little above a two-dollar annual income, would stand condemned in the face of the two-dollar average outgo which so often goes with present high prices for feed. When feed cost but a dollar a year, they would still have been considered profitable. When the cost of keeping doubles, as it has in the past few years, it becomes a necessity to have a bird that can produce a greater income. And it behooves one who would make money from market eggs and poultry to be very careful in the choice of a breed.

Although the Rose Comb Brown Leghorn does not fall strictly in the American class, she is more nearly like the American breeds, such as the Wyandotte, than any other familiar to me and not in the American class. She may almost be called American-made, since our breeders have developed her very largely from the

Single Comb variety, and she is somewhat better suited to our climate. The chicks do better on range than when raised in confinement on bare ground, but that is probably true of the chicks of every variety. The hens, on the contrary, thrive and produce well in confinement, giving eggs high in fertility, that will produce good chicks. They always keep in the handsomest of feather, except during a brief period at the end of the moulting year, and there is no known breed, I think, that will keep in such uniform good condition as to dress, when kept in constant confinement in bare yards. Although the raisers of the large exhibition breeds are loath to admit it, the Rose Comb Brown Leghorn is really a "world-beater."

This variety is one of the most difficult to photograph with good results, for two reasons: the sprightliness of the birds, and the color combination, which is one that does not "take" well. Still, some very fair pictures have been made of the Brownies, by those whose patience is almost limitless. Few others can succeed in this difficult attempt.

IMPROVING THE AMERICAN HEN

Ideals — Government Introduction of new Breeds — Faults of Rocks and Reds — The Wyandotte as a Basis for Improvement

As long as there is any Yankee blood left in living Americans, so long, at least, there will be continued efforts to improve what is even now regarded as almost up to the ideal. If the present "ideal" is reached, all that needs to be done is to set up a higher ideal! Most of the aspiration in late years has looked toward the production of a breed, or strains of existing breeds, that will be worthy the name of "two-hundred-egg hens." The United States Government, in collaboration with the Maine Experiment Station, has been at work along this line for ten consecutive years. In its last report, it announces failure along previous lines of work, but announces also that it will work henceforth along a line of more strict selection of both parents, especially as to ancestry. It proposes to introduce new blood of the same variety into the birds now at the Station, and also to cross birds of other pure breeds on the Station stock of Plymouth Rocks (Barred). Any that show up well "will be retained and bred further."

Between our average production of about seventy-two eggs annually per hen (1900 census), and the two hundred and fifty-one secured from a single very special specimen at the Maine Station during the coöperative work, there is a wide stretch. The problem, as it presents itself at present, is to raise the very low average of seventy-two or less, up toward (to?) the very special two hundred and fifty. The Government has taken as a basis of effort the breed already most satisfactory to the people of this country in general, although it is not the variety having the best average of unusually high reported records. But as yet in ten years "there has been no increase in average egg production." (See 1908 Report.)

According to the present popular demand among American poultry raisers, — among whom the practical class is, of course, in the tremendous majority, — practical points count first. It is asked that the birds which are to distance those already well known shall be, possibly, of a little higher grade of table quality, a little better breeders than some of them now are, and that they shall increase in laying probabilities and actualities just as much as may ultimately be found possible. If it were possible to produce a hen that would lay three hundred and sixty-five eggs a year, I think about ninety-nine out of a hundred poultry keepers would be satisfied!

The general grower of poultry cares little about breeds, new or old, if he can get such a bird as he wants. But the immense quantity of poultry literature, especially in periodical form, which is now being broadcasted throughout the country, leads continually to a better knowledge of existing breeds, existing conditions, and existing desires. One of the desires that is inherent is the desire to see what one can do with the existent, in the way of betterment. And it is always assumed that the resultant of effort will be betterment. More than this; it is almost universally asserted that it *is* betterment, when it may be really the reverse, but giving better results by reason of the special interest taken and special resulting care given.

Among fanciers in general, however, it is very generally believed that the nearly one hundred varieties of hens alone now recognized by the American Standard of Perfection are decidedly more than enough; especially as so small a proportion of them are really entitled to be called popular. About one-third the American varieties, one of the Mediterranean, and one of the English, are about all of the present Standard varieties which can be strictly termed widely popular here. There are, possibly, another dozen of varieties which would keenly resent being left out of the first class, and which have a really strong following among us. But this still leaves more than three-fourths of the present sorts in the list of those we might be about as well off without, were it not for the life which large variety gives to the exhibitions of fancy stock.

It is a fact, however, that in spite of the wide feeling against the desirability of introducing new breeds, our own Government is planning to enter the field of introducing such breeds, provided it finds something really worth while. The production of a breed or variety of more practical value than anything which we now have is the only logical excuse for introducing more breeds. And I have been cogitating somewhat over the question whether it might not be well, in the future, to restrict such output of new breeds entirely to Government workers, and their collaborators. Since every other individual is quite sure to be suspected of having an axe to grind, official recognition by a perfectly disinterested source is the only method that can save the poultry raisers of the country from spending much hard-earned money in finding out how far the claims of any new variety or breed may be sustained in actual trial. For instance, there is just aiming to get itself into the lime-light of general public notice a breed which claims to have "a record" of three hundred eggs per year. In order to find out whether or not these marvellous claims are true, every one interested (and what poultry raiser can help being, if he believes there is a chance of getting a three-hundred-egg hen?) must pay out a snug little amount for trial stock or eggs. It would seem that a breed introduced in the above way might become popular in a much shorter time than if entering commerce after the present fashion; for the Government bulletins go everywhere, and as a rule their brief, unbiassed descriptions carry a conviction which subscription papers and advertisements cannot hope to emulate. The power of an advertisement, as a rule, consists in making a claim to desired characteristics, and repeating that claim till repetition has worn a channel for belief in the reader's consciousness.

The positions of the exploiter of a breed, and the one who has it under consideration with a thought of buying, are in many respects radically different. The owner looks continually at its good qualities, and refuses to admit, even to himself, that it may have serious faults. For how could he carry conviction to the buyer at all, if he believed his stock to be of little value? The considering buyer, on the contrary, wants to know the worst about the variety, and he is likely to discount the owner's statements as to its best. Imitating the prospective buyer, let us look a bit at the various American breeds, and note what points call for betterment, and guess, if we can, what changes the average poultry raiser would try to make, were he aiming to "improve" upon existing American varieties. As a matter of fact, this is just about what every breeder would be doing, if he saw his way clear. And this, despite the almost universal advice from experts to let crossing alone.

Improvement by selection is always legitimate and possible, and it is practised by every breeder or poultry raiser worthy of his opportunities. Crossing pure varieties in order to make others is another matter. Yet it is affirmed that it was deliberate intention and planned crossing to get desired qualities that produced the only English breed that has made any headway against the Amerian sorts in this country. The Black Orpington, thus produced, has many enthusiastic breeders, and, as it is one of the more recent breeds, this is really a very strong argument for trying yet again, in the effort to get a still better one, on much the same line of procedure.

The well-established old Plymouth Rock — what faults prevent it from being thoroughly satisfactory? And is it not straining a point to criticise a breed that has established itself by sheer force of merit from the lakes to the gulf, from one ocean to the other? Most certainly it is not: for even the greatest admirers of this most popular variety will admit, if you catch them at the psychological moment, that the breed has faults. Some of these faults have been partially overcome in some strains by selection. But in the great raft of these birds throughout our land they still exist, and in some they have become intensified. I don't think I ever knew a man of average interest in poultry raising, no matter if he raised Rocks only and could not be induced to raise any other existing breed, who did not grumble about the Rocks because of a few exasperating defects in an otherwise satisfactory fowl. One of these faults is the extreme tendency toward sitting; another, the tendency to lay on internal fat; a third, the slabbiness of the growing chick, after it begins to stretch up.

The demands of the market and the color preferences of individuals work together to narrow the really popular fowls down to those of three colors,—white, red, and buff. White and black in combination has a good following, where the market is not finical, and the Barred Rock has advanced in spite of the overwhelming preference for white. On the farms, where large vermin of all sorts make war upon the poultry, the mixed color of the Barred Rock is one of its strong points. But a black- or other than yellow-legged fowl has small chance for popularity here. Perhaps this has kept the Javas out of the running, since they are universally admitted, where known, to be good birds — many say better than any of the Rocks. The Rhode Island Reds, the Buckeyes, and many varieties of the Wyandottes are not old enough to have fully proved what their place will eventually be, but we can forecast this with fair accuracy through knowing their color. The Buckeyes will make a place for themselves on merit and color, if the breeders do not spoil them in the making. Their defects are: too long legs, too strong a sitting tendency, unreliability as to hardiness. Some types are very hardy; others as far the other way. This delicacy will probably disappear, as their numbers grow greater, color better, and inbreeding less.

In Mr. J. H. Robinson's book, the edition of 1899, now just ten years old, he said of the Rhode Island Reds: "They are only locally popular, but are becoming celebrated for hardiness and prolific laying. In meat qualities they are considered inferior to the other American varieties." We look back in amazement as we realize that it is only ten years since this now well-liked and well-disseminated breed occupied such an uncertain outlook. But the breed has had plenty of "push" behind it; its hardiness and other good qualities have spoken loudly in its favor, and its color is universally admired. That is to say, the color of the best specimens in the shows is admired. In general hands, the breed is sadly mongrel-like in color, only an occasional bird showing the beautiful, ideal red. Even what it has, some say, it got, in part, from the Buckeyes. Perhaps there are few breeds which, being adventured on because of beauty, could prove so very disappointing in average hands. I have almost never seen a novice breeder of Rhode Island

Reds with a decent looking flock. One woman, who bought one hundred eggs of a good breeder, did get a very fair bunch; but the general custom is to buy only a sitting, when making first trial of a breed, and this gives small chance for selection. The usual outcome of this situation is that most of the culls become a part of the new breeding flock, with disastrous results. And such methods affect a comparatively new breed worse than an established one. The chicks of this breed — Rhode Island Red — are such greedy feeders that they are quite subject to indigestion, and a brood sometimes "goes off" almost without warning. The fowls are quite subject to roup, and, it seemed to me, more subject to "going light" than many others. This, too, may possibly be traced to their greedy habits of eating.

The White Rock has made a strong bid for public favor, and those who predict that this variety will in time be the most popular one in the country, have a good basis in the facts. At present the White Wyandotte stands much higher in popularity, and it is the nearest to the American ideal of any bird now known. Ten years ago it was classed, although then still rather new, as the most formidable competitor the Plymouth Rock had ever had to meet. To-day, it stands much higher in popular favor than it did then. As to form and carcass, it is very near our ideal. It is an excellent winter layer, but it does not equal its predecessor, the Silver Wyandotte, as a layer, and it is a rather indifferent breeder. That is, it demands small numbers of females to one male, and it does not hatch as well as one would like it to do, unless on range under the most ideal conditions. The eggs are not quite as large as we would like, nor so uniform in color, although selection is improving them along these lines very rapidly. And, although it is of about the right size, it matures a little too early for soft roasters of good weight. However, the eyes of American breeders are at present strained most intently in the direction of a heavier-laying bird than we have at present, and, if we could get that, even without any other great improvement in the popular American varieties, we would not need to complain much. This is a very strong point, for a poor layer must have a maintenance ration even as a good one, and the increasing price of feeds for some years past makes it seem almost imperative that the income be also increased. All the increase in laying beyond present possibilities in average hands would be so much clear profit the more, after allowance is made for feed.

It has been my conviction for some years that the desired advance must come through the union of the Wyandotte with one of the breeds in which the egg-producing instinct has been strong for many generations. Doubtless the Leghorn is the best example of such a breed. The Brown Leghorn is considered the most widely distributed variety, and there are, I think, few farmyards where traces of it may not be found. The Rose Comb variety matches best with the rose comb of the Wyandotte, and a cross of these two varieties, the Rose Comb Brown on the Wyandotte, perhaps the Buff variety, would be fairly suitable. This cross, bred back from this point on the pure Buff Wyandotte or even in time to white, ought to give us a little nearer the ideal American bird than anything we have at the present time, good as are the present American varieties. At least, if I were to make an attempt to improve the American breeds, something like this would be my mode of procedure. This cross, if selected carefully to Wyandotte type and the solid color, would become virtually an improved Wyandotte, having the additional laying capacity which all desire, and, it may be, a better and a trifle plumper carcass, since the Rose Comb Brown Leghorn is a most excellent market bird, barring the small size and dark pinfeathers.

QUALITY AND NUMBERS AT THE SHOWS

In the "Waste-basket"—Numbers of American Varieties—Comparative Showroom Values—the Newest Breed—Beauty of Feather

AT our fancy poultry exhibitions may be found a sort of waste-basket arrangement known as "the any-other-variety class"—a class despised of exhibitors—into which many a breed or variety, not yet strong enough to command a class, is liable to fall. Many a breed has made its first appearance here. The show catalogues and reports offer a half-reminder of this waste-basket in the way the breeds are handled. Attention is focussed on about a half-dozen varieties, and the rest are treated, in some sort, as "any-other-varieties." All the elaboration, in the reports, all the special attention, is given to less than a half-dozen breeds. A single brief paragraph taken from the report of the New York show for 1909, in one of the foremost poultry papers, will show this tendency accurately: "This show seemed, more than ever, strong in the general-purpose class of fowls. It was the Plymouth Rocks, Wyandottes, Reds, and Orpingtons that furnished the big classes. Many other classes were fairly represented, and quality good, but few were strong in numbers. Barred Plymouth Rocks were the largest class."

If anything were needed to establish the claim of the American hen to consideration, the merest glance at this list would furnish it. For it will be instantly noted that of the four breeds in this bunch of leaders three are in the American class, while the fourth is the strongest competitor of "American" breeds. The contention is proven without effort and without any successful chance to even oppose it. And, while it was affirmed that the Plymouth Rocks led all breeds and varieties, I noticed at the show that the Columbian Wyandottes, though not second in numbers, appeared to be second in attracting and holding general interest. This is a rather remarkable showing for a variety almost new and in a show numbering so many varieties and such large exhibits.

In order to get a little glimpse into the near past, I picked up two catalogues of the New York show in recent years, just the ones I chanced to have at hand. They were for 1904 and 1906. The number of exhibits has run from three thousand up for some time, and the 1906 catalogue numbered fifty-six hundred and eighty-nine exhibits, including pet stock, poultry supplies, etc. But No. 3608 was reached before the lists began on pigeons, which always number largely.

QUALITY AND NUMBERS AT THE SHOWS 81

In 1904 the Wyandottes led, there being nearly six hundred birds of this breed, two hundred of which were in the breeding-pens. The Silvers, Whites, Buffs, Blacks, Silver Pencilleds, Partridges, and Columbians all pushed together to roll up this big exhibit for the breed. The Plymouth Rocks mustered something over four hundred in number, while the Rhode Island Reds had just short of one hundred representatives.

In 1906, the tide of competitive interest was running strong, the Rock breeders had their grit up, and the list of Rocks in all the varieties counted up the astounding number of nine hundred and fifty specimens in this one leading show. Of these, about four hundred and sixty-eight were of the Barred variety, two hundred and sixty-nine of the White variety. The Wyandottes, all told, went nearly to eight hundred specimens, there being three hundred and thirty-nine of the popular White Wyandottes. The Rhode Island Reds were far behind these, though far ahead of most of the other breeds, reaching something over two hundred in number, and the classes generally being quite as large as the ordinary fancier cares to meet. A breed like the Reds, having but two varieties, naturally labors under a disadvantage when breed numbers are counted, as against a breed like the Wyandottes, numbering eight well-liked varieties, or the Rocks, which are getting to be a close second in this line.

G

"Interest" flows in two streams at such an exhibition. One of these is commercial, the other is merely sentimental. Commercial interest may also be divided, since there is the interest which accrues to the good performer and that which accrues to the most successful from the beauty standpoint. The streams meet in that breed or variety which can best illustrate both beauty and performance. At a large show the interest in the Barred Rock classes and the White Rock classes is very strong, but these are far more nearly "fanciers'" breeds than is the White Wyandotte, for instance, because, back of the fancy quality, — perhaps dominating it, — is the enormous strength of the exceptional power as a producer of eggs, and as a producer of the ideal market carcass, which belongs to the White Wyandotte. I noted, in a show report from Madison Square Garden in 1909, the statement that the Columbian Wyandottes were "next in interest to most poultrymen present" to the largest class in the show, though much farther behind in numbers, — as so new a variety would naturally be. I think it is because this variety combines in one so abundantly both the streams of commercial interest, besides being a good example of the beauty which attracts the mere sentimental interest of the casual visitor, that the Columbian Wyandotte has already taken so high a place in general public interest. This has been put into a prophecy by Mr. J. H. Robinson, an editor who keeps his finger continually on the pulse of the poultry interest, in these words, delivered in February, 1909: "We are inclined to think now that if the breeders of White Wyandottes don't look out, this variety (Columbian Wyandottes) will lead the Wyandottes within ten years."

The White Plymouth Rock has owed some of its advancement to the mere fact that it was a Rock, and thus a member of a breed already widely popular as a general-purpose fowl. It owes some of it to the fact that it is white, and thus makes a nice-looking market bird. But as a fancier's bird, I think it has owed most to the fact that it was a whiter bird than the White Wyandotte, its strongest rival. It has been said that the bleaching of white birds "is killing the interest in white varieties." It is said that this is carried to the point of illegitimate faking, and this to such an extent that the fact that a bird is white in the showroom is no criterion that it is naturally an exceptionally white bird. And as the big prizes turn very largely on exceptional whiteness, as exhibited, this has been termed "the worst form of faking."

It is, then, no longer safe, no matter what it may have been in the past, to buy, as a breeder, an unknown bird which has won a high prize. Thus, the very foundations of the fancy poultry business have been cut under. And from this point of view the big fight at the last meeting of the American Poultry Association as to comparative handicapping of the white and black varieties becomes rather a farce.

In the matter of beauty, the American breeds have a good lead, since they include the laced, the pencilled, and the brilliant red types, which are the very foundations of the beauty possibilities. Take away the red color, the pencilling, and the lacing, and you have small material for making a beauty breed, as things now are. There is a growing feeling that breeds to be admitted in the future must be able to show actual distinctiveness.

One new breed, admitted but two months before this writing, made claim not only to distinctiveness rather unusual, but also to beauty of a very unusual degree. I am not sure that popular opinion will uphold all this claim. Personally, I do not admire the breed extravagantly. But its combination of colors, chestnut-red and clear white, is rather striking, and, though it shows the well-known laced plumage, the lacing, as shown, was beautifully clean and notably wide and even in all sections. The bird is to be known as the "White-laced Red Cornish." This is a brown egg breed, claimed to equal the Leghorn in egg production, and to lay as well in confinement as on free range. I think the breed has virtually no literature at the present time, but a number of judges spoke in its favor at the convention meeting where it was presented for admission to the Standard of Perfection. To the eye uninfluenced by previous prejudices, these birds appear rather as a variety of Cornish Indians (once known as Indian Games) than as a distinct breed. But they are claimed to be in every sense "a twentieth-century fowl," and though they may not fall into the "American" class, they are an American production, and, in one sense, can claim to belong to the group known as the American hens.

The lacing and pencilling of exhibition birds would be a matter for amazement as to its beauty and its exactness, were it not for the fact that we are so familiar with it. In the American class we have at least six varieties which have attracted prominent notice to an extent to put them into the Standard, which show the striking lacing or pencilling. We have also the barring of the oldest form of the Rocks, and we have two exceedingly handsome Red breeds, in three varieties. We have plain black, plain white, and the Light Brahma combination of these. Thus it appears that the birds of the American class lack very little that would fit them to form a whole show of their own if need were.

THE AMERICAN FARMER, THE AMERICAN HEN, AND THE FANCY

The American Standard — What constitutes Perfection? — Overdoing the "block" Idea — Ideals — Behind the Farmer, the Fancier — Good Business Capacity a Necessity to the Fancier

ALTHOUGH the American hen has been evolved, in the last analysis, with a view to adjusting her exactly to the needs of the American farmer, he is still far from being fully acquainted with her. There are two chief sources of information as to her fancy standing, these being the Standard of Perfection, and the poultry show. The Standard of Perfection has just been subjected to revision by the American Poultry Association, and will from 1910 onward differ in some points from its present form. But in the large essentials it will be much the same as before.

As our farming people become yearly more interested in pure bred poultry, many of them looking to become breeders of fancy stock, the interest in The Standard, by which all such fowls are judged, becomes yearly more widespread. Hundreds would like to know just what the scope of this Standard is.

Many would like to breed by The Standard, yet feel its purchase to be an unjustified expense. These are much mistaken; the man who would breed fancy fowls needs, first of all, "something to go by"; this is precisely the use of The Standard of Perfection to the breeder. And if the breeder must follow its laws, so, too, must the judges; the friction and disagreement of which mutterings are sometimes heard arise from varying interpretations of The Standard's text. And, although one might think this unnecessary, it is really very near an impossibility to paint such a word picture of any object that all readers thereof shall see the object alike.

If critics are disposed to criticise The Standard as not clear, they may do the public a great favor by explaining what The Standard means to say, in terms which are understandable. The new edition will, it is hoped, be beyond criticism. Surely no volume needs more to be exact and correct in its English.

The special text for each variety includes the special disqualifications which would hinder it from competing in any show, detailed description of the color of each section of the male, and such description for the female also. Interpretation seems to vary more here than elsewhere, as various individuals seem to have different color values as a mental standard. The "rich, golden buff" of The Standard means one shade to one judge or breeder and a different shade to another. The judge who is not a breeder (or who has not bred considerably the variety which he passes upon) is not worth very big pay; yet if he is a practised breeder, he is sure to have a standard of his own, which he supposes to be *the* standard, yet which differs from that which plenty of other honest breeders believe to be the standard.

In spite of the efforts of years, and periodical revisions in the interest of advancement and accuracy of statement, the novice breeder who for the first few times tries to apply The Standard to his birds, to see how near it they really come, is rather apt to be filled with disappointment, if not disgust. He had expected to be independent, with the aid of The Standard of Perfection. He finds, at once, that he must formulate for himself a second standard, a sort of sliding scale of comparison between the average of birds as they are and the book standard. He must know perfection in a bird; he must then know how near perfection the best specimens grown will come, and also how far below these all allowable breeding and exhibition birds will fall. This can be learned in part, slowly and laboriously, in his own breeding yards, if he have good stock; but the only reasonably swift way of learning it is to visit those shows which employ judges who apply The Standard sternly, see what wins the various prizes, and talk with the judges, if possible. This private standard is not an absolute one, by any means, but one comes gradually through all these means of instruction and education to sense what a good bird, a fine bird, or a superfine bird is. He learns that even the last is not perfection; this being at first a hard saying.

Glancing over the law for shape in Plymouth Rock males, I see that the comb is to be medium or slightly below medium in size, that the ear lobes are to be medium in size, and that this happy "medium" is called for fourteen times more in the description, besides two demands for "moderate" development. Some standard of comparison must be in the mind of the learner before he can grasp the meaning of this word "medium." He must know something about small, medium, and large breeds. Thus it appears that some general knowledge of breeds is necessary to quick comprehension of Standard terms of description. Here, again, we may see how attendance at a well-judged show of high character is a help in learning breed shibboleths. Any poultry show is helpful, but a good show is very much more helpful than a poor or even a "medium" good one, as so many poor birds will appear at a lower class show. The novice, as a rule, believes all show birds to represent high value, and is thus apt to be confused in his standards of comparison. One of the first things that strikes the learner, at even the big shows, is the number of birds which are very poor indeed in some prominent section. It is, of course, against the law of good breeding to use such birds as breeders, and against the law of good exhibition to place them on show in competition; yet the superior value of such birds in other sections is expected to carry the fault through. It often does it, even against the better judgment of the judge then acting, and against the general spirit of breeding and judging law. At one state show I saw blue ribbons on the coops of two different birds with combs unspeakably poor; and, as I walked the Pan-American alleys with a novice, I noticed his chief comment was on the surprisingly poor sections in a good many of the birds, even there.

The Leghorn is a long-established breed; but though The Standard said for years, "tail carried well up, but not upright," there is even yet question as to how a Leghorn tail must be carried to be just right. This is largely because just as soon as the breeder gets the tails low enough to avoid "squirrel tail" all the time, he finds the tail inclined to get too low. If he aims to get them "well up" he soon has a goodly (or ungoodly) proportion that are not only upright, but in front of the vertical line. So, even in the best shows, the best birds will show tails a trifle too long, too high, too low, too something nearly every time. The owner of a first-prize bird in New York told me that the judge had scored his bird there, and asserted that it deserved a full score of ninety-eight. Happy man, who stands alone in attainment! And happy the novice who could form his idea of Leghorn perfection from this exceptional bird!

The man who would breed fancy fowls for high prices is most unwise to try to do this nice work without all the tools necessary. Chief among these is The Standard of Perfection. Yet with one tool alone, he will be too heavily handicapped. Let him get every good tool he can acquire, and he will not then be so well fitted as he could wish. Let him study books, men, and birds, if he would get on in this chosen work. A talent for asking questions is a great thing for him who knows how to season all replies with the proper salt. In questioning a lot of men, the replies of some will give a pretty fair idea as to the amount of salt needed to mix with those of others, so that even the learner has his compensations and need not be thrown too far off the track by any fanatic. All this is in the line of acquiring experience, and experience always counts.

What constitutes perfection in a fowl? The American Standard of Perfection — always the latest edition! — is, broadly speaking, the answer to this question, as its descriptive matter is the law for one hundred and thirty varieties of domestic fowls, both land and water classes. There are those who maintain that the word "perfection" is a misnomer, since perfection could not improve; while the Standard is being changed with a view to improvement once every five years. It is perhaps a fair statement to say that the Standard is revised by all interested breeders, since at the last revision copies of the changes proposed by the special Revision Committee were mailed to all members of the American Poultry Association, long enough before the latter's annual meeting for them to be fully conversant with these proposed changes, and fully prepared to discuss them in open meeting, and to vote on them at the final call.

The hundreds of changes last made, while important, were, as a rule, not radical, the intention being to raise the Standard of each breed as fast as it shows susceptibility to advance. As we have fowls of many shades of color, all through buffs, whites, reds, and combinations of black and white, as well as plain black and plain white, it takes many color terms to describe them. One important, if minor, part of the last Revision Committee's work was to reduce the number of such terms, yet still retain enough to describe clearly every shade and combination known. The number has been reduced about two-thirds, so that the new Standard will contain thirty-five of such color terms.

The American Standard of Perfection further describes itself on the title-page, as "a complete description of all recognized varieties of fowls." This means not only those of the purely American classes, but also those of other countries in their "American form" — if I may coin an expression. The public is forbidden to reproduce its pictures or to duplicate its text, except in brief quotations for the dissemination of knowledge. The first edition of this American Standard was published in 1874, and all changes have been based on this work, new breed standards being added as they were recognized; admission to the Standard constituting this recognition. Some breeds which have been hindered in various ways from getting early recognition have gained a degree of public recognition which enabled them to dispense with the formal recognition of the Association fairly well, and mutinies and rumors of mutinies among the specialty clubs have been known at times. If a breed have a strong club, it is able to defy the Association to some extent; but all recognize that it is to the interest of any breed to gain this formal recognition, if at all possible.

One of the first things to appear in the Standard is an illustrated glossary of several pages of technical terms. Understanding of these is necessary to intelligent reading of the body of the Standard. There is a list of general disqualifications and an explanation of the various "cuts" which judges are required to make for defects of all the various kinds which may detract from the value of a "Standard-bred bird." Any bird of pure blood which is bred with an effort to make it conform as far as possible to the excellencies described in the Standard of Perfection, is termed a "Standard-bred" fowl. This is one of the most satisfactory words to use in speaking of fancy fowls, as there is no uniformity of use of the terms "thoroughbred," etc.

The Standard describes birds of English origin, those of French, Italian, Asiatic, etc., but always from the American point of view. Older Standards are usually the basis of our descriptions, no matter how much these may differ after we have worked our will with them. After we have bred any fowl for a time, American ideals appear to influence the breeders and the breed, so that the American type of any breed is seldom a close copy of its type in its own country. Even the Orpington, — a late English fowl, — as we breed it, grows to differ quite materially from the Orpington of the English breeder's ideal form. The Orpington ideal picture, from the pencil of England's best living poultry artist, shows a bird of strong Cochin type. To be sure, the cushion does not rise to a height greater than that of the main tail-feathers, but it is there, and at least on a level with the upper tail feathers, in the female. To be sure, the breast feathers do not sweep the ground, as they virtually do in our Cochin ideal, but they swing down so low that they give much of the same impression, and the bird looks loose and saggy all over. The American ideal of the correct Orpington, as shown in our Standard, is as close-hauled as a Rock, the level carriage being the most noticeable distinction on a quick glance. There is more difference between the English and American types, as thus pictured, than there is between some distinct breeds. Still another variation in type may be seen in the illustrations in some American poultry papers. One of the present year, by one of our best artists, and depicting a winner of this year in one of our eastern shows, has much of the English looseness of feather, with a shape never before, I think, seen in any bird. (The depth of body immediately back of the leg is almost exactly fifteen-sixteenths the length just a little below the tail, while in our Standard "ideal" the depth is a little more than two-thirds the length at that point.) This bird won first over a number of others ranking as first-class, and was hailed as among the best representatives of "true Orpington type" now existing. The typical Wyandotte, heretofore depicted in our Standard by our best artist, is barely five-sixths as deep as she is long. Thus, this representative Orpington is shorter, in proportion, than our Standard Wyandottes. Yet the same publication has recently published an appeal from Mr. Felch entering a plea for greater length of body in the Wyandotte!

Beside this fact we might place the statement of a certain Wyandotte breeder who claims to have raised his flock to an annual average of one hundred and ninety-seven eggs. This gentleman affirms that if the craze for short, blocky specimens does not stop, it will utterly and surely ruin the Wyandotte as a utility fowl — by which he means, as the context shows, an egg producer. "The more blocky you get them, the poorer layers you are making of them." The depth of his best layer, as pictured, is practically two-thirds her length, measured on the same plan as the previously mentioned specimens.

With as many breeds as we now have, it is a very difficult matter to present ideal pictures, even when based on photographs, that will show at a glance the real difference between the breed types. This statement might easily stir up a swarm of denials; but I saw a pretty good illustration of the fact at the annual meeting of the authoritative body of this country, as regards poultry. Artists of ability presented some specimens of half-tone work, intended for the new Standard. The work in itself was beautiful, but the breeders there present tore it to tatters, metaphorically speaking, and the Association was all of a buzz over it. And I think this has been the case with a large proportion of the pictures presented for criticism, even by our best artists, whose ability no one questions.

The forward step in the effort to get good Standard illustrating is to be "idealized photographs," in half-tone work. Indeed, the pictures discussed above were of this class. But breeds and breed types so overlap each other, and the different poses of any particular specimen make so many seemingly different birds of her that the artists have an almost impossible task. In the Red breeds, another difficulty appears: The shading necessary to bring out proper form approaches its deepest at the points where the living bird is its lightest, in some parts — a mechanical and artistic problem which would puzzle the wisest, since the result tends to deceive the very beginners whom the illustrations are especially made to help.

But how will this help the poultry-raising farmer? Not long ago, a farmer wrote his favorite paper concerning a poultry show which he had just attended. His comment was that from first to last it was purely a fancier's show, the farmer and his needs being wholly ignored. I quote: "I talked freely with several visitors whom I met at the show about its characteristics, its limitations, and its shortcomings, and we all agreed that the exhibition was of no practical benefit whatever to the farmer; that, beyond this, it was of no benefit to the public generally, since it in no way tended to improve or increase the supply of commercial eggs or dressed fowl; and that, therefore, from both the farmer's and the public's point of view such exhibitions were not deserving of public support."

I think this critic rather shallow, but he is one of thousands who have no conception of the way in which the fancier—berated individual—stands back of the farmer and his utility work. There is a great cordon of workers continually hammering at every problem in poultrydom, of whom the general utility man knows almost absolutely nothing.

The oldest representative of the poultry papers of this country is about forty years old; its nearest competitor is, say, thirty. Within a space of forty years, then, the present immense interest in poultry has been "worked up." Do you object to this term, and affirm, rather resentfully, that poultry which pays is certain to come to the fore without any effort to "work it up"? So thought Micawber, in effect, while he waited for something to turn up; so thinks the business man whose competitor works all around him, and secures all the business, while the first waits for business to come to him!

No; all these things influence each the other, and all help the industry and advance prices.

There are two distinct lines of work with poultry,

known as the "fancy" and the "utility" branches. That is, utility tries to convince itself that they are distinct; "fancy" knows that it has no enduring, wide-reaching life without its coadjutor, "utility." Take this instance: A fair Single-comb White Leghorn breeding cockerel is freely offered at $1.50. Can one buy an equally good White Wyandotte for that amount? No; because the utility value of the latter is about twice that of the Leghorn, so that a Wyandotte of about the same grade as the Leghorn in question will command about $2.50 to $3. The breeder will simply say that he can't afford to bother to ship a cheaper Wyandotte; better sell them to the butcher outright and be done with it, for there is nothing in the price margin to pay for coops, extra care, and advertising. Thus, utility bolsters fancy prices.

When "utility" believes and affirms that it could get along as well, if not better, without the fancier, it forgets that the fancier made the breeds which we now have, and brought them to their present perfection. It forgets that virtually all of the men and women who are dispensing poultry instruction are fanciers, in greater or less degree. Something over fifty years ago market chicks ranged from two and one-half to five pounds each, the latter being thought very large. Now, since the fancier has manufactured the wonderful American breeds from the scrawny dunghill and the barrel-tall

Shanghai, there is scarcely a grower but has owned or seen a Plymouth Rock female, either white or barred, weighing nine or ten, or even eleven pounds, when alive. Now the city markets often have to refuse stock because it is too large! Yet with all this it requires the laws of "The Standard of Perfection" and the great and constant effort of the fancier to keep his stock up to weight, in order to insure that the stock on the general farm shall come up to common market requirements in weight, so rapidly does deterioration in size and quality go on under average treatment. The probability is that the simple habit of hatching late chicks and using the later ones for breeding stock is responsible for the larger part of this deterioration in size; the ravages of lice unfought would account reasonably for the rest, without any need to seek deep-lying reasons. Yet, such is human nature that, were it not for his positive Standard, even the average fancier would let down the rules enough to cause retrogression.

Each year hundreds of poultry shows are scheduled. The first lone progenitor of all these is reported to have been held here in 1848. Between that time and the present, the education of the individual has been progressing. I think it may be safe to say that by far the larger part of those who now rank as fanciers were first inoculated with the poultry fever virus at a show of fancy fowls. Some may have had literature first, especially in the departments of the farm papers, but the usual line is, I think, first the show, with its revelations of possibilities, then the desire for fuller information, creating the demand for more and better literature. Literature and shows have had such an effect in increasing genuine fanciers that good fowls, from The Standard point of view, can be found in almost every town of any size. The show in the modest town often brings out some birds about as good as any that grace the big shows. The fancier has certainly given us handsome and quite uniform flocks of birds.

He has given us the handsome American breeds of middle weight, plump birds, fine for market. He gave us rather early the Americanized Leghorn, unapproached by any other style of Leghorn. Since that triumph, he has had a difficult task in trying to surpass himself. A bird with the egg capacity of the Leghorn and the market value of the best bird one can imagine is what he has since been striving to produce. When he almost reaches this ideal he carefully adds just one more touch of Leghorn blood, and behold his bird is once more a bit too small — too near a Leghorn, in fact — too near the Leghorn type. She has hard flesh and small carcass, with too much neck and leg, and not high enough egg power; between extremes the worker in breeds has been seesawing for years. When he reaches the mean, his bird is too general-purposey; she isn't q-u-i-t-e what we want for market, and she does not lay q-u-i-t-e so many eggs as the Leghorn. And there he stays. The reproductive faculty is a sensitive one to deal with, and man has been aiming to get the hen back to the fish stage of unnumbered progeny. We see advertisements of 200, 210, 213, and 242 egg "strains," but, between ourselves, we don't believe in them. Meanwhile, one poultry journal of highest esteem throws doubt on the attainment of the two-hundred-egg hen, even while another blatantly affirms that she has already arrived. And still eggs — just common eggs — are so strongly impregnated with the money flavor at midwinter that most people cannot eat them at all! Indeed, I know people who produce them who cannot eat them on this account. But if, when the hen reaches the fish stage of almost unlimited production, the eggs shall also have a fishy odor, what better will it be? Shall we not be still unsatisfied and demand of the fancier that he make us something nearer our taste?

The question as to the value of fancy stock to the farmer would be answered in the negative much oftener if it related to fowls than if it related to other farm stock. Yet, if fancy fowls are admitted to be in any way more valuable than common stock, it ought to be answered in the *affirmative* oftenest when it relates to fowls, because fowls can soonest be made to return a yield from the investment.

THE FARMER, THE HEN, THE FANCY

Perhaps we do not always realize the many good points which poultry and poultry products possess. The late Mr. Tillinghast, whose clear common sense was such a help to the Rhode Island Station farm work, used to call attention to the fact that poultry gave quick returns, with products done up in small "packages," easily handled, easily marketed, small in compass compared with value; and, in case of eggs, capable of storage for a considerable period in the producer's hands. These points are well worth consideration.

It becomes every farmer to inquire whether fancy fowls have any points of value above common fowls that will justify the extra cost to him of getting a start with them. I believe it to be useless for any producer to pay out money for extremely high-priced fancy stock *unless* it is to be cared for in a businesslike way. It will take only one or two years to spoil any flock of fowls, no matter how valuable it may have been, unless a man will care for this as he would for any other investment.

Again, if a man is going into fancy poultry with a view to becoming a "fancier" in the process of time, he might just as well "quit before he begins," unless he is a good business man and a good salesman. For, after the stock has been carefully raised and intelligently selected, it still has to be sold. There are only two main ways of selling it: through advertising it in the press and through advertising it in poultry shows. Who is going to buy your stuff unless you have some way of proving that it is worth buying? A man wants value received (some want more), whether it be plums, pigs, poultry, or any other product of his hand and brain, added to Nature's forces.

It seems to me a marvellously interesting and glorious occupation to join one's own skill and intelligence to Nature's generous and beneficent agencies, and to become in a limited sense a creator of improved individuals in all lines of effort in two kingdoms — the vegetable and the animal. Nothing can exceed this work in its delight.

As to the fancier, why say he is but a devotee to certain ideals of beauty, and therefore his work and his products have small value to the practical man, who wants above all else to see dollars grow? Is this quite true? If the fancier confine himself to externals only, like the color of feathers and the shape and length of wattles, it may be a just way to view his work from the superior viewpoint of the practical man. These may be his chief delight, but the fancier is far more practical than the man who looks down on him, for he gives nearly one-third of the one hundred points that go to make a perfect, ideal fowl to shape of the various sections; and, if he demands certain shapes of comb, lobes, etc., he also demands specific shapes for breast, back, body, etc., and these in the general-purpose fowls, are what give the practical value many times. The scorner may say, "What nonsense, to demand that a fowl shall stand with legs wide apart!" But is it nonsense to insist that a fowl shall have a broad breast? The fowl that has legs placed wide apart has a thick breast between them, and here is where the fancier takes care of the interests of the practical man. In a large number of varieties, the back must be broad to conform to the Standard, and this helps keep the rounded body that makes a sightly market specimen. The fact that the fancier breeds every specimen with a view to its conforming as nearly as possible to the one ideal form for its breed makes for uniformity of the finished specimens, and here again the fancier is the salvation " of the practical man." For it is in this desired uniformity of product that the great value of fancy fowls to the average farmer lies. The more nearly even in size and shape the specimens of dressed poultry which a farm turns out, the better position will the products of that farm take in open market or in the estimation of special private custom.

With eggs the story is still the same. It pays the handlers in the large cities to sort eggs to color. Why then should it not pay the producer to do these simple things which add to the value of a costly product through only a little care? The easiest place to do this sorting is in the hatching eggs, because this soon secures something like uniformity, not only in a special lot, but in the whole output. The fancier makes this selection in the hatching eggs to certain extent. The Standard does not demand it, but his sense of beauty — *and* his customers — often demand it of him. The point to be made by the practical man against the fancier is simply that the fancier cannot be as severe in his culling to practical points as he would like, except where his standard is sharply explicit; because he must also consider surface points and must give much attention to color. But, though he may be tempted to give undue attention to color while his breed is new or while he is working up to high-grade stock, there usually comes a time when he has the color question well in hand and can then give a better proportion of attention to the more practical points.

The practical man who does not care to be a fancier does not himself need the high-priced birds, but he makes a mistake when he thinks the fancier's work is nothing to him. His very best hold is to build his work on a fancier's fowl that has many practical points, because he can work from this along the practical line without the handicaps of the fancier. He can buy the shape and size and uniformity cheaply. The uniformity which the fancier has established can become his by the expenditures of two dollars or three dollars for pure-bred eggs, and if, from this selected stock, he will himself select for uniformity in size of carcass and for uniformity in shape, color, and size of eggs, it need be only two or three years before he can have almost his own ideal in poultry products. He can get along faster than the fancier does because he has only one or two things to handle, and these do not contradict each other.

A great reason why this condition does not come about on the farm is that there is no standard of practical points to live up to, except in the mind of producer or buyer. A standard having authority to demand certain practical points would be a great help. The trouble is that in the press of other things, men do not hold themselves to the mental standards which they may have set up. Conditions and circumstances continually tend to draw us all away from our standards, if these are not fixed in some way. It takes excellent business ability to raise and sell fancy stock and eggs, and make the occupation, as a business, bring in more than goes out, because the expenses are always considerable and the details many. Advertising, circulars, and the like make expenses count up pretty well.

It takes good business ability even to hold a lot of fancy stock at as good a point as when first acquired. It has long been the custom for fanciers to advise farmers to buy cockerels of purebred stock to "grade up their flocks." I do not think this amounts to so very much, unless it is carried systematically through several generations. If good males are secured, their blood worked into the flock, and their progeny mated back to them, progress will be made. But in this case I would advise that either the best layers or the most uniform females, as to both size and color, be first selected. If specimens which answer both these requirements can be found, so much the better; and if trying to grade up a flock in this way, it would probably be best of all to use only one female the first year, the very best one in the flock. The second year would then give one a good flock, all descended from the very best.

There is one valuable point in fancy stock which is not often referred to in farm papers, and I think many do not think of it in connection with fowls at least. It is really one of the very strongest reasons for introducing the blood of fancy fowls to the farmyard. This is the fact — a fact more true of the oldest breeds than of the newer ones — that purebred stock will reproduce itself more surely, not only in externals, but in its possibilities for production. That is, an extra layer of a pure breed is much more likely to produce good layers than is a specially good layer of mongrel blood. All the characteristics which have been bred into a purebred fowl are strengthened so that it is "in the blood," as we say, to do certain things, to reproduce certain qualities. This is a modified meaning of the "prepotency" so often mentioned in connection with larger stock.

The experimenters to whom we look for advancement along the lines of better breeding of plants and animals think they have discovered the law on which these things work out. If they really have the A B C of such a law, we may rest assured that they will soon be able to give us D and E and F, and we may look forward with confidence to the day when we may receive from them even the X, Y, and Z of these hard breeding problems.

THE AMERICAN HEN AND AMERICAN MONEY

Their Relations — Our Poultry Products — Money Out — Beauty vs. Utility — Getting at the Dollars — The Market Bird and the Dollars

THE question as to the American hen's actual relation to American money is still, to some, an open one. Does she absorb more good American money than she produces, or not? As an argument against the hen, it would be very effective to quote a recent delivery of the Government, to the effect that it is probable that "more failures are made in poultry farming than in any other type of farming undertaken by beginners." On the other hand, the same authority avers that it is "decidedly one of the best and most profitable types of farming, when properly conducted." This, it will be noted, throws the credit for success, or the onus of failure, directly upon the owner, rather than upon the fowls. This means that the American hen is all right, in the estimation of Uncle Sam. It is the Yankee himself who is on trial.

It is to be hoped, and it now seems probable, that we shall have more reliable statistics in connection with the next census than have ever before been possible. The American Poultry Association has tendered its services in helping the Government to get a more accurate report of things as they really are. It passed a resolution characterizing former methods as "inadequate," and urging that the next census be taken "in a manner that shall best show the kind, quantity, and true value of poultry and poultry products on the farms and in the villages of the United States." Few persons have understood that the Census Bureau has been handicapped in this matter, because it has had no authority from Congress to collect statistics of poultry, except as these were incidental to a census of "the farm." This arbitrary distinction has necessarily thrown out of the records, all but those poultry establishments which could be considered, either in fact or by assumption or courtesy, as "farms." The Bureau decided that the income from the plant must count, gross, into the hundreds of dollars, or the work must take a considerable part of the time of one person, in order to come within the limits of "farm" production, in the census meaning. There is little doubt that the coming census, even if very imperfect, will be of wider grasp than any previous one, as lists of large poultry places are to be advertised for by the Department, and individuals have already been asked to keep accounts in order to be ready for the enumerator, with facts, rather than estimates.

Under such difficult conditions as have formerly prevailed, we have had a report of $281,178,247 as the value of poultry and poultry products produced on the "farms" of this country in 1899, the year previous to the last census taking. In between the ten-year periods, we get estimates, based on the wide knowledge which the Department always has of existing conditions and of continuous progress. For 1905, its estimate was half a billion dollars as the value of the poultry and eggs produced. There are those closely in touch with poultry production who have ventured to predict, that the next census returns will total "close to a billion dollars." I suppose we may then talk about "the billion-dollar hen" and put the now famous "ten-thousand-dollar hen" out of the public eye completely!

There are statistics already in hand to prove that a goodly gain will be made in the figures, because of the increase in selling prices alone. Many have come to believe, and are saying freely, that farm products will never again be low-priced, in this country. Should this prove true, the " wealth" of the poultry farmer is sure to increase; since, so long as he continues to grow his own grains and to go without hired help, he has the game, to a certain extent, in his own hands. The product is one which, in some lines, has no competitors. Can another farm product be named with such a stable demand, so rich in nutriment, put up in such convenient packages, so easy to carry to market, and extending its period of income-production throughout, or nearly throughout the year? The American hen is a pretty safe "proposition," be assured.

As to actual increase in price, the figures of some large consuming points are available. In New York City, the average wholesale price of dressed poultry has advanced since 1899 well toward two and a half cents a pound, as recorded up to the year 1908. This one fact affects more farmers than any other that could be named, except change in egg production, since poultry raising is a branch of farm industry probably found on more farms than any other one branch.

The rise in the price of eggs has been so gradual, much of the time, yet so continuous, that we are almost amazed as we look back to see from what we have advanced. Possibly St. Louis has ranked as one of the lowest markets for this class of products, and New York one of the highest. Fourteen years ago, the lowest price of eggs in St. Louis was eight cents, and for seven years it ranged from six to nine cents a dozen. The extreme high price in those years ranged month by month from nineteen to twenty-five cents. Since that time, it has ranged in December and January from twenty-one to twenty-nine cents, touching twenty-one and twenty-two only twice, as a monthly mean of the higher quotations. In New York City, in 1907, the high average was twenty-five cents. In 1907 and 1908, the low monthly average touched twenty-four or better, six different months — the colder ones, of course. The high average, which I understand as meaning the average quotations for the fancy grade, touched points between forty-four and fifty-five, inclusive, six times. These are the wholesale prices, it is to be noted. If the value of the eggs on so-called "farms" amounted ten years ago to more than one hundred and forty-four million dollars, it is easy to see that advanced prices alone would now lift the values high above the figures then collected.

The average farm price of eggs for the entire country was, in 1899, 11.15 cents; by 1904 it had risen to above seventeen cents; in 1908 it was 18.3 cents per dozen. In the advance sheets for 1909, price-movement, rather than actual prices, is noted; but as the advance is given as being more than 100 per cent over the period just before 1900, we can see just about where the figures will fall.

It is well known that fanciers — who are quite largely town poultrymen — make a business of importing stock and eggs from foreign countries to improve their flocks when the stock of those countries ranks higher than our own. But this summer I have known of several instances of genuine farm fancy-poultry breeders turning to imported stock to get just what they want. When a good Yankee reaches the point where he will not even wink when abstracting from his pocket a roll of twenty-five dollars or fifty dollars for a single bird which he thinks may help him to win a first, or to improve his flocks and make from them many winners, it begins to look as though there was money put into poultry, whatever came out! Last spring, too, I received the supposably modest circular of a couple who were just beginning on a farm of their own, and was much edified to learn the correspondingly modest prices for their eggs for hatching. The lowest, if my memory serves, were five dollars; the really "good" ones, from twelve to twenty dollars per sitting. And the situation seemed to smack of irony, when a few days ago I read, from the pen of the foremost English authority, the statement that their particular variety is of so little repute in its native country that it "has no history." If this is strictly true, it must be that we are making an American hen out of this variety by American breeding. And I might add, just as an item of additional interest, that it is to this variety that the "ten-thousand-dollar hen," owned in this country, and exploited throughout the world, belongs.

The crux of the whole matter is in the uncertainty as to whether our good American money is to appear eventually on the side of profit, or that of loss — whether it shall stick to our pockets, or to those of the other fellow. It is my opinion, largely speaking, that the matter of selecting the American hen, her of the strictly American class — to speak Standard-wise — may be of great importance just here. My pronounced belief is that the hen of the American class is the one most likely to help the good American money to settle finally to rest in the pockets of her producers and handlers if they are average people.

The amount of money that goes into advertising, in our very numerous poultry papers, — not to mention the far more widely circulated farm papers, — into expressage and exhibition entry fees, into club publications and dues, etc., is almost beyond belief. When one goes to the New York show and is met by the simple news item that X sold fifteen birds to Y yesterday for fifteen hundred dollars, one feels no earthquake, for such items are becoming, if not common, at least nearly enough common to be among the expected things. One may go home and muse on the character of the situation, should the fifteen contract roup in the showroom or get smothered in the express cars; for these incidents are rather more common than are those meaty sales. But they are all in the game, and if one hundred dollars or five hundred dollars of good American money goes out of existence with a roupy hen, it need not cause a ripple. There is more money to put into hens, and usually there are more hens.

At least, while beauty is thus bringing in the dollars, plodding utility, likewise beautiful, it may be, is bringing in the pennies. Four cents for every pearly white offering of Madam Utility, or for every brown-tinged, Bostonese variation in the egg story, is — catch your breath! — twelve dollars a year in gross income from the eggs of that "recorded" three-hundred-egg hen. The man who has a thousand of her kind, is assured of an income of twelve thousand dollars. That most precious article of henhood cannot, to save her precious substitute for a soul, eat much more, on the average, than the commonest cull from the egg-fold, so that feed, even in high-price periods, cannot cost more than about a sixth of this income. And there you are, and the American hen has made you, in the course of a few years, a real aristocrat! Who says that figures will lie? They will not; they will only leave out a few of the particulars.

There are various ways of getting at the dollars, one of which is in operation near New York to-day. It is a rather new thing to poultry literature, no matter how old in actual existence. It consists of a system of branch egg-farms, circled about a main farm, located within city limits, producing chicks in large numbers from the best stock and after the most approved methods, to sell to the branches. These branches raise the chicks, produce from them infertile eggs, and ship these eggs every day or two to the main farm, which sells all the eggs to private city trade. This main farm has a delivery system whereby all eggs go to consumers at once on their arrival from the branches. It is a great idea: the system is its own producer, its own middleman, and its own retailer. How could the problem of the middle-man be better solved? The twenty-eight farms in this system, at the present stage of the work, consider themselves able to handle seven-thousand layers. If these were three-hundred-egg hens — but what is the use? They are shelling out the dollars, and supporting the American farmer, which is honor enough.

If there is one characteristic which, more than others, marks American farmers as a class, it is their appreciation of the utility side. And it is the American farmer, more than the man of any other class, who should be interested in the American hen; who is, indeed, interested in her. Hence, it goes without saying that the discussion of her from the utility standpoint ought to touch him to the quick, even though not so sensational as the showroom story.

There are three special points in connection with the utility side of the American hen, viz., the winter eggs she can be made to lay, the kind of a market bird she makes, and the proportion of her chicks that can be raised under average conditions. The American hen as a winter layer is not to be named in the same class with others, because she so far outclasses them all. It is not denied that the Mediterraneans often make good winter layers, but it is at the cost of warm houses, close confinement, and the acquisition, many times, of the vices which are a concomitant of this treatment. It is not denied that some of the Asiatics make very good winter layers, but they must needs be hatched in bitter weather (some hatch them in January for next winter's laying), and they must be fed three or four months longer than the Mediterraneans or the American breeds before they are even ready to begin laying. It is probably quite within the truth to say that two Leghorns or Wyandottes can be raised to laying maturity on what it would cost to raise one Asiatic to the same period of maturity. In addition to this the Asiatic will afterwards take the room of two Leghorns, and almost the feed of two. At least, two of the large breeds will eat regularly as much as three Wyandottes or Reds. The man with an eye for the utility points will not be slow to figure on these items, and to decide that the middle-weight, middle-sized bird, with enough of the Asiatic about her make-up to insure her winter laying qualities, is the bird to depend on for profit. It hardly needs to be said, perhaps, that whether or not she pays a satisfactory profit in winter will depend very largely on her handling, as the simple fact that she has the desired tendencies will not insure her being profitable. The feed and the housing are both essential points, and are so crucially important that they will be discussed in another chapter.

It is quite well known in this country, — at least among those who take special interest in poultry — that the demand of American markets differs considerably from that of, for instance, the English markets. The white-skinned bird which our English cousins consider the only one to furnish chicken meat fit to eat is the very one which would lie unsold in a large proportion of our markets. It has been said that this condition is changing to some extent. Personally I have seen no proof of such change. The yellow-skinned bird is the American ideal, and the American type — which virtually means the general-purpose bird with yellow skin — has largely driven all others out of the general markets. No doubt there are very many birds of other breeds sold in our wide country, but a very large number of them go to local markets, and have no influence on the general markets at all. The men who raise stock chiefly for market in this country are very likely to choose the White Wyandotte or the Plymouth Rock, except for a few special markets which may like larger birds. The average consumer has an average family, and wants an average sized bird as a meal for that family. He does not want a bird the size of a pigeon, nor can he manage one of the size of a turkey, like some of the Asiatic males. A certain eastern commission firm, describing the class of market stuff desired, laid stress upon the birds' being well fatted "so that the breast bone does not stick out," on plumpness, and on "yellow meat." This gives a raiser his cue at once, if he knows anything about the various breeds. The middle-class or general-purpose breeds are the only ones that can fill this demand with profit to the producer at all times. The Wyandotte has the plumpness desired, even without fattening, as has also the Rose-comb Brown Leghorn. This Leghorn makes an extra fine broiler, but the American breeds are better for market stock of the roaster class.

A market fowl should have the capacity to take on fat readily, and to keep in good condition all the time. Probably the Rock takes on fat most easily, but the Wyandotte has the better breed form, being short-necked, round and plump in natural conformation. Many like the Red breeds, because the under-color and "pins" are also reddish, and the pins that may remain after dressing hardly show against the skin.

I fancy, from what I have seen, that many people have little conception of what a really fat market specimen is. Such might find help in the display of market poultry at the larger shows in this country. The good farm wife would be likely to call them "butter-balls," for they are butter-colored, rounded so that the natural form scarcely shows, and look like parcels of fat only — so greasy is the skin. I think these are often a bit overdone for the sake of emphasis, and of winning the prize; at least I have never seen such fat birds on actual sale in any market. But they are a good object lesson.

The American varieties have another point of fitness, in that their period of attaining maturity is so reasonably short that they can be adapted to the selling season far more easily than can the large breeds. The man who has had his novitiate avoids, as far as possible, putting his stuff on the general markets at any time between early September and the middle of January, because this is the season of lowest prices. Since, during the first months, the cost of production is measurably the same for the same age, in average conditions every cent of drop per pound in price must come out of the profits — the *profits*, I say; that margin between cost and selling price which ought to get into the producer's pocket. If he sell after prices drop, he sells to his own loss nearly always, since drop in prices is usually more rapid than possible increase in weight, even with growing birds. This point emphatically helps to determine where the dollars will settle.

The American breeds can easily be brought to salable condition and size before the drop in price for early chickens. And the "fricassee" bird has her innings here, also. In surburban towns there is a constant demand for what are known to the buyers as "fricassee chickens." These are hens preferably a year old, and the call for them continues at least until the time for roasters at a reasonable price. The American breeds give a bird of popular size, and one usually in fair condition. This opening allows a man who keeps fowls for eggs chiefly to market his hens all through the summer, as they pass the productive period for the year. Since such a grower will want to replace at least half his hens with pullets, he can furnish just what custom asks and at a time to suit his own needs; which, surely, is all that should be desired.

MAKING A HIGHLY PROFITABLE WORKER

Investing too Eagerly — Vigor Necessary to Continuous Work — Weeding out Incapables — Errors in Selling — Preserving Profit-margins

If one who knows nothing at all about poultry raising begins to spend money for eggs, fowls, appliances, etc., it needs no Solomon to be able to predict that the money in poultry for this person will be the money put in for some time to come, unless he move very carefully. This is just where we of the farms have two or three prime advantages. We generally know something at least of the actual work; we have a native caution, as a rule, that hinders us from making too heavy investments where there is uncertainty; and a very large number of us are kept from mistakes we might otherwise make by the very fact that we haven't the ready money to risk. Money in hand is a bad thing ofttimes for the man with a new craze, the attractions of which may be laid before him in the guise of figures which always show money coming into his pocket. He has only to hold it open!

A year or two ago two people — pretty intelligent

people, too — came into a poultry supply store with a long list of stuff which they wanted to buy on the spot. They were of the earning class, bitten with a belief that they could make money out of poultry faster than they could earn it by their usual hard work, and full of eagerness to invest quite a pocketful which they had saved from their earnings in all sorts of things wherewith to make a "start." It chanced that the salesman had been through the mill, and was a man with a conscience. He sold them an incubator, a brooder, and a few other necessary things, then said to them: "Now, see here; these other things all down this list are not strictly necessary. You can get along without them at first or you can make things that won't cost you much take their places. You're going to need a contingent fund for emergency calls which you can't see ahead now. Better save the money you meant for these things till next season, anyhow, then see how you feel about investing in them."

Those who took a superficial view of the situation would have said at once: "This man was hired to sell all he could to customers who came to the store? What moral right had he to put these people off the track and lose his employer money, even to save it to the intending buyer, poor though the buyer's judgment was?" Events, however, justified the man completely. The same customers came back next year, inquired for this special salesman, thanked him for saving them from throwing away money and became regular, intelligent, continuous patrons of the house which had, possibly, saved them to the ranks of poultry keepers by making their first year one of reasonable income rather than one of unreasonable and discouraging outgo.

I have known of several cases in which people have sunk all the spare cash they could lay hands on for several years in buying one thing after another for carrying on their poultry fad, and have finally dropped the matter disgustedly, calling down anathemas on all who might mention poultry keeping at all favorably. This is the group of whom one may always safely predict failure from the first minute. I have known of three deserted poultry places, standing object lessons and arguments in their community against the "ridiculous" idea that poultry keeping may be profitable. In one community I knew of numbers of successful poultrymen whose work could not offset in the public mind the effect of those deserted, pathetic, tumbling buildings. Yet in nearly twenty years' acquaintance with suburban towns about New York I have never known more than the three instances of which I speak of failure at all notable.

Just recently this matter was brought freshly to my mind by a new recruit inquiring of his favorite expert where a poultry plant already established could be had for rent or sale near one of our large cities. The inquirer made this statement: "I keep seeing that many people have started in with too large places, and are obliged to sell at a very low figure, but I am unable to find any such places to buy." A part of the reply was, "Just at present I know of none." Before we bought our present place we got out a search warrant, as the saying goes, and covered New Jersey all about New York, and we were just as successful as the inquirer above. Scarcely anything was to be found in the scores of suburban towns lying thickly about the big city.

Last week I was looking over the prospectus of one of the college correspondence courses in poultry. In it the statement was made that positions could easily be found for qualified students. Indeed, the greater trouble was to supply desired help, since so large a proportion of the students preferred to go into the business for themselves. These are of the class who already know the risks and the expenses. The case has stood thus ever since the college poultry courses began. And it seems to me no two better arguments than the above are needed or could be had to prove that poultry is giving satisfactory returns to those who are in the work.

The facts noted above ought to be especially helpful to the very large number who do not quite trust the hen, — who fear that anything at all expended on her may be money thrown away. Few days pass that I do not have queries of some kind to answer. I am obliged to look at the matter from the standpoint of the man too eager to spend money — in fact, eager to throw away money — and also of the one too fearful to be reasonably generous or liberal. There is nothing like the logic of facts, and when I have these to throw light on any question asked I am satisfied, because I feel certain that I shall not unwittingly mislead any one.

The other morning the milkman came in rather excitedly. One can easily see that a milkman can add eggs to his stock in trade without extra effort, and make a fine thing of a private custom. So, the one of whom I speak keeps several hundred hens. Even so, at this season he cannot begin to keep even with the demand, and he buys all the eggs he can find from desirable sources.

"And what d'ye think pullets brought at the auction yesterday?" he began. "I went over there, thinkin' I c'd perhaps find a nice lot of pullets. What d'ye guess they brought? Sixty-five cents for them, not half-grown. Ninety cents for some they called Minorcas — I don't know if they was pure, but I don't think so. And $1.09 for Plymouth Rocks!" "Were they pure?" I inquired; "if so, that was not exorbitant for good pullets." "I don't think they was pure. And hens brought eighty cents."

There is no place in the country, I presume, unless we except Boston, where so much money is put into poultry as in the surburban places tributary to New York City. Feed is high and supplies are bought to a large extent in the big city, where high rents, cartage, expressage, etc., add enormously to cost. The trade in poultry supplies runs into the thousands of dollars weekly during the season. Lumber is high, help is high, and the suburbanite who would dabble in poultry must put his hand deep into his pocket and pull out the money to put into poultry, even though he has cut off every useless expense. Under these circumstances, to have people crowd each other to buy pullets as we near the winter season at prices double what they have cost to raise, seems to speak rather loudly in favor of poultry as a means of profit even when the money put in must be the largest sums usually considered within limit.

Remembering that the hen, to be highly profitable, must be a consistent and pretty continuous worker, we ought to grasp easily the importance of constitutional vigor. Yet it is notably the fact that wherever I have been among the farms of different states there seems to be a failure to appreciate the difference between fowls of fine constitution and those of a little less than average vigor. No doubt this is because the conditions in general assure a fairly good average constitution. The fact that birds running at large are not likely to be so closely observed as are those in confinement also plays a part.

One who raises fancy stock soon learns to rate a good constitution higher than any other one feature, except in case of fine exhibition birds. If he wants to buy a bird for a breeder, one of his requirements, particularly noted, is sure to be "Must be vigorous." Sometimes it will read, "Must be exceptionally vigorous"; and it is well for every one who handles domestic fowls, either for utility only or as extra fancy stock, to learn not only to distinguish between the bird of fine constitution and the one with a weak hold on life, but also to harden his heart against giving the latter "the benefit of the doubt." *Culling is a chief factor for success.*

While it is true that accident and circumstance will be pretty sure to occasion losses in any flock, no matter how well handled, careful weeding, with the thought of the requirement, "Must be vigorous," in mind, will reduce these occasional losses to a minimum. Lack of care in culling, especially for vigor, is the producing cause of a very large proportion of the losses among flocks in general. The judge in a show of fancy fowls is instructed, in case of doubt, to give the specimen the benefit of the doubt. But this is a fatal position for one to take who is selecting specimens for utility or breeding purposes. We are all pretty much alike in this — we hate to cut out a tree well started, no matter how much damage it is doing; and we hate to condemn a bird that may improve, because we do not fully "sense" the consequences of her remaining in the flock.

Any one who will notice the condition of the birds that "take" every disease going and hand it on to better ones, will find that such specimens have been off in condition or vigor in some way. It is taken too much for granted that the disease has "run them down"; but if they are examined closely at the very beginning of the difficulty, they are almost sure to be found already off condition. What birds are they which first succumb to roup? Those which are underlings, or weakened in some way, or dissatisfied with life. What chicks are those which are first overcome with vermin? The weakly — says a poultry raiser of fifty years' standing — those which have not vigor enough to rid themselves of parasites. What hens are those which break down under ovarian difficulties? Too often those whose stomachs and livers cannot do the work required for such a busy egg machine.

The practical point is, How is any one to go to work to weed out the incapables? By what token may one know that there is trouble ahead in the flocks or for certain individuals therein? Simply listlessness, more or less pronounced. A poultry editor not long ago asserted with some positiveness that life isn't worth living. Perhaps it isn't for a man who is off the track; at any rate, I've seen many a hen that silently agreed with him. Like man, the first symptom she shows is lack of interest in what has hitherto been the chief item of existence, the tri-daily meal-time. Her opinion as to the utter stupidity of life is expressed in demeanor as clearly as man himself would express it, were he speechless.

Except in the matter of diseases like roup, following severe colds, the ailments of fowls are almost wholly in two classes: digestive or pertaining to the reproductive organs.

Nature's imperative law is that the fit survive the unfit; but Nature takes her own time and her own way to end the story. "Going light," liver congestion and its long-drawn pining, ovarian inflammation with its consequent barrenness and languid letting go of life, are her chief variations with regard to fowls. All of these are ailments which keep the average owner guessing and which leave the ailing birds a long time in the flocks as a menace to their betters; since any epidemic or acute trouble which comes along is pretty sure to seize on these first, and they transmit the contagion.

The other day I received a little book which, it seems to me, has not a little bearing on this question. Perhaps readers will be amused when I say that it was a brochure from one of the leading sanitariums of the country — not a sanitarium for languid hens, but for people who have partially let go their hold on the life so dear to most of us. In this little book the statement was made that more and more is the conviction forced upon experimenters that all disease is, at the bottom, a matter of food and digestion. Given good food that can be digested, and necessary fresh air, and we get good blood, which gives health to every organ. Well, this is pretty near what has been the theory, if not quite so strongly stated heretofore, for long years. But these experts base their assertions on the results of treatment of thousands of sick folk, and upon actual experiment with the contents of the stomachs of hundreds of those under treatment, said contents being withdrawn for examination. Verily the stomach tube is a boon to those of us who don't have to make its acquaintance, and the physiologies may now give poor Alexis St. Martin (who had a hole in his stomach wall which made him a public blessing, because he was so good a subject for experiment) a well-earned rest! The most important thing learned in these researches, perhaps, is that various foods, which may be classed, have power to stimulate the acid-producing glands, or to quiet them. This means much to our nation of dyspeptics. The great point is that these people are putting less and less faith in medicines, more and more in fresh air and proper food, and are curing a larger proportion of patients each year. This makes the matter of suitable, nourishing food, even for hens, seem a more important one, but I want especially to put the emphasis on the necessity of watching the eating habits of the birds. Any single member that shows no eagerness for her ration, any slow eater, any bird that gets in a corner by herself, is a suspect, and, in all probability, is not worth the room she occupies. I don't mean, of course, that a single instance proves anything. I do mean that listlessness, wherever seen, is a challenge to open eyes.

There is a possible exception to some of the above, in the case of later hatched chicks yarded with a lot of more mature ones. If kept there, they promise to prove worthless, but if placed by themselves they may prove all right. A large proportion of the failures of beginners lies in the fact that they do not keep young chicks separate from older ones. I once saw a man who was planning to teach all his predecessors how to handle poultry, place all his young stock at midsummer in close yards with his old stock, which had already been confined there a year, and which had only room enough before the advent of the chicks. The second year thereafter he was "out" of chickens, and the old standbys were still plodding on in their old ways.

Yet a man new to handling chickens often makes a greater success of it than the old hand, if he has plenty of hard sense. A man came to me to talk chickens, who had just moved out from the city and knew literally nothing about handling them. He had a man on the place who had some experience. The owner wanted the fowls which he had just bought shut up. The helper objected. But the city man said sensibly: "I can see no reason why they will not do as well yarded from now on (as there is practically nothing they can get outside), provided that we furnish them with sufficient exercise and green food."

There is one point in handling the birds in winter that receives too little thought in many quarters. This is the providing of separate quarters, if at all possible, for the pullets. I do not believe so strongly as some do in the superior value of any pullet over a well-moulted yearling hen; but I know that the hens will frequently make life a burden to the pullets nearly all winter. They will drive them from the feed, drive them from the water, drive them off the roost, and then pick them on general principles. And the pullets cannot grow or produce at their best under these conditions.

Some one who wrote me not long ago concerning some little trouble in the flock spoke of about thirty-five hens, I think, and a dozen or so roosters kept with them, to kill along through the winter. So many cockerels as this, unless very immature, would cause more loss than people usually imagine. Besides, they made the flock larger than is desirable for best returns. They could have been cooped quite closely elsewhere, and would doubtless have grown faster than they did as they were handled. Even pullets of two or more differing breeds, like Leghorns and others of a heavier breed, do not prosper quite as well together as if all were more nearly alike. And where most of the lot are feathered alike, one or two birds widely different in feather are often made the butts of the entire flock. I dare say many will think things cannot be drawn as fine as this, under average farm conditions, but it is comparatively easy to avoid such circumstances, if one looks ahead enough.

The thorough housewife has a saying that bread, no matter how well and carefully made, may be utterly "spoiled in the baking." In like manner the output of a vigorous, working hen may be made of little profit by errors in selling either eggs or market stock. And it is a nice matter to plan to catch the market just right. I might mention an instance, quite typical: —

With the rapid fluctuation of prices and the sometimes inexplicable fads of the broiler demand, it becomes a very nice question just when to catch the market for broilers. Prices vary surprisingly in different parts of the country, but even in the same town management may make it true that one poultry raiser will receive far larger returns than fall to his less lucky (?) neighbor with the same market chances.

One year we had a few chicks hatched in late March,

which brought at the door, alive, over fifty cents apiece in June. Yet they were not early enough to catch the highest prices. The next year we purposed to do better, and began hatching the first week in March, instead of the last. The early chicks did exceedingly well, and the same buyer spoke for them who bought from us the previous year. Depending on his promise to call for them in a few days, we sought no other market. Nearly a month passed, and when the tardy buyer finally appeared, the chicks which were scarcely up to desired weight a month previous, were too large for the market. They could not be used. Many of the second hatch were also beyond the market demand, and much of his supply must come from the third incubator hatch. One cannot nap much where the broiler market is concerned.

The reason for this finicky attitude of the buyers seems hard to find. That is, it appears so to the average seller, who does not consider the source of the broiler demand. Much of the call comes from high class boarding houses, restaurants, and hotels, which serve one-half a chicken as a portion to the table customer. Of course, too large a chick costs the proprietor too much, as he charges no more for it than for a medium sized one. This limits the broiler call to two pounds live weight or less. All larger stock must go on into the season to sell as "roasters" after broilers are no longer attainable in goodly quantity.

The question is, must we lose money on these birds that pass the broiler stage? That will depend on how fast the price goes down, and to a considerable extent on the breed. This point does not always receive sufficient consideration. Up to broiler age, say, one and one-half pound alive, a Leghorn will grow nearly as well as a Wyandotte or any of the larger breeds, and a cross of Leghorn on the larger breeds makes a very good stand-by for a quick growing, thrifty broiler. But at just about this period the Leghorn begins to fall behind. Two months later, the bird of the larger breed will weigh one-half more and will have eaten little, if any, more. And this question of number of pounds plays an important part in that profit, after the price per pound has ceased to be of such paramount importance that nothing else is considered.

The pound and a half broiler would have brought in June forty-five cents, alive. If late in August he weighs three and a half pounds and brings twenty cents a pound, dressed, he will be worth seventy cents; if he weighs five pounds, dressed, at twenty cents he will bring one dollar; in either case, it will have cost something like twenty cents for extra feed, leaving from ten cents to thirty-five cents to pay for waiting on him two months, dressing him, selling him to private custom, delivering him, besides the risk, which, between cats, dogs, owls, crows, thieves, and disease, is no small item.

It is this large risk which really makes poultry raising so — well, so risky!

Still, generally speaking, this margin will not make it a bad bargain for good handlers to keep the broilers for later roasters, unless their room chances to be a great deal better than their presence. In cases where prices for roasters drop much lower, or where the birds are thoughtlessly kept till the holidays, to sell almost surely on a low market, the profit on them would probably be less than on the broilers, especially of the small breeds. Indeed, it would be difficult to sell the cockerels at all, later, if hatched very early, as they would, if Mediterraneans, be so nearly matured as to come into the hated "rooster" grade, in the city market, down in the five and six cents a pound column.

It amounts to this: If you are raising many early chicks, in order to get the precious early pullets that make the early winter layers, there needs to be the utmost care that the cockerels are disposed of as early as the market will take them. Many of those who raise Leghorns for layers especially have been fond of saying that the sale of the broilers just about paid for their own raising and that of the pullets. But to make this come true, the cockerels must be disposed of at just the right period, when the profit over their own cost will be largest. The first cost of early hatched chicks — that is, the eggs, oil, attendance, etc., needed to produce them from the eggs, before attempting to raise them — is so heavy that no chances must be taken, if there is to be a real profit. Some people assume a profit where none exists by ignoring first cost and interest on investments.

The poultry raiser needs always to recall that there are "two ends to a ladder." The layer herself must be of the right type to fit the conditions. She must be vigorous, a good feeder, a good exerciser, a bird with some brains as well. But to make her highly profitable, the owner's work and brains must be linked with hers. It is as much "team work" as some athletics.

EGGS OF THE AMERICAN BREEDS

A Standard for Eggs — Brown Eggs more Abundant — Ideal Shape — The Brown-egg Breeds — Large Eggs — Cold Storage — Fertile and Infertile — Almost Perfect Fertility Possible — Losses through Infertiles

OUGHT a variety to be judged by its eggs, to any degree, beyond exacting that a popular sort must lay many of them? If so, why is the one law for American breeders silent on the matter? Is it practicable, indeed, to give utterance in the Standard to any law concerning eggs? These are certainly fair questions, and when raised they demand a fair answer. There are many seemingly desirable things which the American Poultry Association has found it impracticable to try to bring about. Since a fowl on exhibition must be judged by what she appears to be, it seems impossible to demand that her eggs shall play any part in her own judging. Doubtless no way could be found to make sure that any particular lot of eggs was laid by any one individual. But a breed standard for eggs, whereby one breed might be judged in comparison with other breeds, might be more nearly feasible.

In New England, where from very force of circumstance they take a more lively interest in poultry and poultry products than is shown in any other part of the country of equal area, a standard for judging eggs was formulated some years ago with which many would be inclined to quarrel. This Standard, while allowing thirty points for shape and forty points for color, gave but fifteen points to weight, the other fifteen going to "condition," which included freshness and cleanliness when the eggs were exhibited. The shape called for was oval, and the color for brown eggs "very dark brown." In my opinion, Boston worked that "very dark brown" idea quite into the ground. In the eyes of the majority of people a "brown" egg is yellowish brown, or near a salmon color, possibly, though on the brownish side. In Boston, for a time, they strove for eggs almost as dark a brown as one could imagine. I submit that this exceedingly dark brown is not an attractive color for an egg. Commission men doing large businesses in certain markets have stated that brown eggs always sell better, and average considerably larger than the white eggs. Whether they were preferred or not would not make much difference just now, for they are offered to the market, which, except in the flush season of early spring, must take what it can get. It is quite probable that the very common idea that brown eggs are "richer" than white eggs, has to do with this preference, as expressed by the commission men. There is no authority whatever for such a notion, as

EGGS OF THE AMERICAN BREEDS 143

authoritative chemical analysis has fully shown. But the rich color of the shell and the richer color of yolk in a brown egg affect the judgment of the buyer, through his eyes, to such an extent that no amount of argument will overcome the preference for brown eggs, except in markets which have been trained to prefer white ones. Since it is claimed that the Barred Rocks are the most widely bred variety of fowls in the world, bred by a greater number of persons and in greater numbers than any other variety, and since they are supplemented — if other breeders will accept this term — by the other varieties of Rocks, and by Wyandottes, Reds, and other American birds, besides the Asiatics, the overwhelming majority of the brown eggs must be assured.

In shape, these brown eggs do run mainly toward an oval, and sometimes an oval so short as to be almost round, rather than to the shape which has given us the term "egg-shaped" — a shape smaller at one end than at the other. While it is possible to over-emphasize this, the ovoid shape is certainly the most desirable shape for the egg, and this shape or any desired shape can be fixed in a few years by persistent selection. It is because breeders are pulled this way and that by the opposing demands of breeding to fancy standards, and selecting to practical standards, that we do not make more progress.

The Rock lays a good-sized brown egg, of which she seldom needs to feel ashamed. The Wyandotte has improved in size of egg, and is capable of still further improvement. The Columbian, being the newest, probably needs the most selection. The Brahma part of her ancestry is surety for her capacity for producing the best, in due time. The Buckeyes and Rhode Island Reds have the capacity, I think, for producing the very finest, most attractive egg imaginable, the pinky brown which accords with their general coloring. Not all of them, by any means, do, at present, produce such an egg. But, since it is in line with their coloring throughout, and since some do now produce it, I think there is no room for doubt that these birds can easily, by selection, be brought to produce this ideal egg.

Not one of these brown-egg breeds can be named with the Minorca, when it comes to size of eggs. But any of them is easily capable of producing for us an egg averaging fully up to — or even above — the common standard of a pound and a half to the dozen. There is no reason, beyond controlling his market, why any producer should desire to furnish an egg very much above what the general market demands. It is not desirable for a medium-sized bird to lay an abnormally large egg, and the egg, as at present known in its averages, furnishes as much *concentrated* nutri-

EGGS OF THE AMERICAN BREEDS

ment, in its one handy package, as the average person not a hard worker at manual tasks ought to eat at one time. I know a man, not a worker at severe labor, who often eats four eggs at a meal. The egg is so proportionally high in protein that this gives him more protein in eggs alone than he should have for his entire meal. Of course, he "complains" much — has digestive difficulties, etc. How could it be otherwise, when a single egg gives one-third the entire protein needed for a meal?

The business producer wants, of course, to make all he can out of his products, and one of the lines along which he has been "speirin'" is that of cold storage. We might think cold storage wholly an evolvement from the Yankee — notably the Boston — mind, if we did not open our eyes to what some others are doing. Over in New Zealand, which wants to export large quantities of stuff to England, the Department of Agriculture has taken a hand by appointing receiving ports, where the Government will take chickens from its shippers, kill, dress, pack, freeze, and cold storage them, for about eight cents each — less than it would cost the individual to get it done independently. Birds are inspected before handling, and rejected, if not up to the standard to which all such stuff is required to conform. Good quality, cleanliness, and promptness of handling are thus insured.

I wonder how many know that it is estimated that from 70 per cent to 90 per cent of all the poultry produced in this country is held in cold storage for a greater or less time. Some of it is carried as long as two years, and some refrigeration companies have reported carrying undrawn poultry "successfully" for four years. Our own Department of Agriculture is on the trail of the cold storage people. Investigation has shown that, although the handlers claim that the stock does not deteriorate essentially, this claim is not wholly sustained, especially for the longer periods. The Department planned, some time ago, for a series of rather elaborate investigations, in connection with the cold storage people themselves.

Perhaps the most important thing for the average buyer to know about cold storage chickens is that, when thawed, they are often soaked for hours, in water of moderate temperature. Naturally, after such treatment, they will decompose rapidly; hence, stock suspected of being thus handled, should not be held after it reaches the consumer. Since it is said to be "almost a matter of routine that every chicken intended for market should sojourn there (in the cold storage rooms) for a certain, or rather, an uncertain time" (see the Year-book sent out by our Government) suspicion needs to be rampant.

There is a legitimate use of cold storage that might

easily redound to the benefit of the producer, near the large cities, at least. The refrigerating plants take in large quantities of eggs, mainly during the season of lowest prices, and this is a reliable means of relieving the market of surplus goods, and carrying them to the periods of scarcity. It helps to insure that the large markets shall have a steady supply of meat and other products of a perishable nature. The producer of eggs in greater number than his local market will take and remain firm, may find an outlet in the cold-storage rooms, where his product may be held till he can handle it, or for some months, until it has advanced notably in price. The charges are not exorbitant, and many producers should take advantage of this outlet. There is no other method of handling eggs so satisfactory, in the long run, as selling when strictly fresh, to private custom. Where markets are good, this method should always have precedence. But where customers are not willing to pay fair prices in the flush season, it is well to hold the product for better times.

I know a certain very small village, which is the outlet for a back-country apple-growing district. In furtherance of its interests, a Board of Trade has recently been formed, more immediately to install a chemical cold-storage plant. This place is not on a main line of railroad, nor is it within forty miles of any large city. Few towns might, with more reason, say, "Where have *we* the strength to carry through such an enterprise?" Yet it is stated that all present storage capacity is filled to overflowing, and the new venture will meet a demand already in evidence. Here is distinctive business enterprise. Fruit-growers are learning that they must use the assistance of cold storage nearer home. The larger poultry producers might take a lesson in modern ways of enhancing the value of their products, during at least a portion of the year, and of compelling the nimble dollar into their own pockets.

Partly because our American fowls have some Asiatic blood, partly, perhaps, because breeds have been newer, with booms, at times, which called for more products than could be supplied in the best quality, the eggs of the American varieties have not always hatched as well as those of the Mediterraneans. The losses, in this country, through infertile eggs alone are tremendous. I do not wish to be understood that the American varieties are notably undesirable in this respect; but the large majority of the birds kept in this country being of the American breeds probably makes it true that these breeds bear a large portion of this loss.

There is a point of view, often urged with reason, which argues strongly for the infertile egg as being of more value by far than the fertile one — aside from the hatching demand — because of its superior keeping qualities. And it is urged that care to produce only infertile eggs for sale on the food market would save to the country the very large losses (sometimes 50 per cent on shipped summer eggs, before they reach the retailer at all) which result from the spoiling of eggs while on the devious road that leads from producer to consumer. Incidentally it is argued that the betterment in average quality of farm eggs by this simple precaution, would make for better, more stable prices. Whether this be true or not, the mind of the thrifty person must revolt at such a loss of good food, especially when most of it is entirely unnecessary.

Concerning eggs for incubation, it is probably quite well within the truth to say that one-third of all eggs set, the country through, prove infertile or at the least unhatchable. Allowing this to be the case, we find that there must be not less than sixty-four and one-half millions of infertile eggs almost entirely waste every year. At but fifteen cents a dozen the value of these infertiles is more than eight hundred thousand dollars. Some fanciers and a very few of the general run of poultry raisers have learned to save some of this waste by feeding the infertiles to the young when first hatched. But even this is only a makeshift, for eggs are worth but a trifle more than poultry meat to feed, or, perhaps, three cents a pound. Medium-sized eggs run about eight to the pound, making the value of eggs for feed less than five cents a dozen.

Many a business under the strict principles of trade makes about all its profits out of its "waste products," it is said. But here is an avocation (which is often made a genuine business, too) in which there is ignorant or careless waste of thousands of dollars of value. It is too often supposed to be a necessary part of the expense; this is true only in a very limited sense. Some infertiles there will always be, no doubt. But it is certain that knowledge and care could save most of this loss.

As the farm is the great poultry rearing arena, and also the place where most of the loss occurs, it is properly the place to begin the reform. The great error made on the average farm is in taking the eggs for hatching from the general flock, instead of selecting the best hens — only a few — and yarding them. It is rather simple to make sure that all of this small number are layers of fertile eggs, while it would be a virtual impossibility to test the entire flock. A dozen advantages follow this method of selection, while the disadvantages are but one or two. Of course there is the expense of yarding, but that will be saved the first year. The other difficulty is that abundant green food must be furnished to breeders, or the chief object will be frustrated.

EGGS OF THE AMERICAN BREEDS 151

With a good vigorous cock (the best fighter in the lot, if he is otherwise good) and selected, tested hens, the remaining points of care are feed, water, shell material, and exercise. If an egg is not perfect, it cannot by any jugglery be made to produce a healthy chick. Perfect eggs call for good feed and water, good digestion in the hen, and shell material, besides the exercise that promotes health. I do not know that we have any patent on producing fertile eggs, but our eggs have been for several seasons so far above the average of fertility that it would seem that our methods were about right. Those methods comprise the points named and a somewhat generous use of meat in the ration provided. This is one of the essentials. It can be over-done, of course, if enough is fed to derange digestion.

Probably the fancier is the one to suffer most annoyance from infertiles, his losses being so much larger on each brood. They are larger, not only because the eggs are many times more valuable, but also because many fanciers do not *provide* the conditions needed to produce eggs of good average fertility. The annoyance, the loss of time for the sitters, the loss of money in the eggs unhatched, the unexpected smallness of the midsummer flocks because the eggs turned out so poorly, make up a total not pleasant to contemplate.

The trap nests which have sprung into such sudden

popularity are, once secured, the best help toward a knowledge of each individual hen's fitness for the breeding yard. The test of advance incubation for a few days only will show which eggs have the greatest average fertility. It is not seldom the case that the eggs of an excellent layer are either permanently or temporarily infertile. When this is the case, the better layer she is the lower will be the percentage of fertility in the flock of which she remains a member, unless her eggs are known and removed. In a flock of ten, one extra-good layer, if her eggs are not fertile, may alone push the percentage considerably below ninety. This matter needs attention, if we would save the tremendous wastes mentioned earlier. This one hen, if her duplicate exist in each breeding yard of ten hens the country through, means a loss, in aggregate, of well toward two hundred thousand dollars.

It must not be forgotten that breeds vary in the matter of fertile eggs, largely in the ratio of their activity. This is the case also with individual birds; hence it is the part of wisdom to select cocks of medium size for their breed, and notable in their breed for individual vigor and activity. The number of females in the yard must vary with the breed. A fair number is better than too few. If the "pen" is small, the cock may run with other birds one day out of three.

A method for the cure of this difficulty that has been adopted in England shows plainly enough that it is there regarded as a matter of care, or the reverse. This method is to replace, free of charge, all infertiles that are proved to come from eggs sold from any individual fancier's yards. When he knows that any possible loss from this cause will come out of his pocket instead of that of his customer, the percentage of fertility takes a sudden and almost unaccountable rise. Precautions are, of course, taken to assure the breeder that it is the eggs sold, and not substitutes, that are returned.

Many eggs are mistakenly supposed infertile, when they are really chilled. Both breeder and buyer need to exercise care in connection with this. Eggs must be gathered often in cold weather, and never should a sitter then receive more than eleven eggs, while nine would be better. Later in the season more can be allowed, but experience says not over thirteen even then for best results. Attention to the points noted on the part of both fancier and farmer will make a wonderful difference in the year's report.

We have received eggs from fanciers who knew their business, which, after being carried hundreds of miles, hatched 93 per cent. In one instance, every egg but one in two sittings was fertile. We cannot expect every fertile egg to give us a strong chick, as an occasional germ will die. But the rule is that the higher the percentage of fertility, the better the hatch not only, but also the better the percentage of chicks that are strong enough to weather the ills of chickendom.

AVERAGE LAYERS AND THREE-HUNDRED-EGG HENS

What makes the Best Layer? — Good Digestion the Key — Some Phenomenal Records — The Munchausen Limit

A STATEMENT recently made by an independent thinker and experienced poultryman is put in a way to stimulate thought in every worker interested in increasing the laying capacity of the poultry flock: "The great layer is best explained as an individual in which systematic effort to increase egg production has reached a limit." This is linked with the affirmation that the hen with a big record in her pullet year is not so likely to produce heavy layers as a hen that is a more moderate layer. If this is admitted to be proved — and the experience of many breeders upholds it — would it not be wiser in the long run for the general worker to cease the continuous effort to increase this one characteristic directly and put his strength on increasing the output by increasing the constitutional vigor of the entire flock?

It is quite generally admitted that the hen with the best appetite and the best digestion is likely to make the best layer, other things being equal. It is more than suspected that a good part of the unusual success with new stock on trial lies in the extra care which it is likely to receive. It is well known that in most cases a flock under average handling will "run out" after a few years. This must mean that the handling under which it exists is not what it ought to be for thrift and vigor. Handling includes living conditions, breeding conditions, feed, water, care, etc.

Just over the way, for instance, is a new poultry-house in which the fowls are not thriving, and the owners, who have kept fowls before, are wondering why. The whole story — as it is now summer — may be told in the brief phrase, "Lack of shade and air." The place they left was an old one, with normal conditions of shade, etc., and this phase of handling has not had to be considered. So, on most farms, the abundance of shade permits the birds to care for themselves in this direction, as long as they are free to choose. But suppose that a mother hen is cooped with her brood in full sunshine, with a water dish but an inch or so deep, and attended to three times a day only, or left to the care of careless children. She is bound to suffer greatly from heat and thirst and likely to lose a good proportion of her brood, no matter how well she is fed.

My neighbor is now on a new place without trees.

His new poultry-house, built at considerable expense by a high-priced carpenter, stands in full, open sunshine, and often with the doors closed. With all window openings, except the closed glass ones, permanently covered with cotton cloth, and with all these openings on the south, the house is an oven, in which the fowls receive daily roastings while still wearing their feathers. It is really a house built for cold weather, and there is no natural provision for protecting it in summer; while the owners are not awake to the necessity of artificial protection. This is only one item. Admit a similar lack of adjustment of thought all along the line, and it will be seen that the hen stands a rather poor chance of being able to do her best, in many hands.

On the other hand, give her the best of housing, the best of feed, the best of all around conditions, then throw out from her number any that show signs of not responding properly to this care. Continue this high average treatment from year to year, and the flock can scarcely fail to increase in average output. This selection may be helped to quite a degree by throwing out all but the best eggs when selecting for hatching.

It takes all sorts of people with all sorts of notions, derived, it may be assumed, from experience, testimony, or imagination, to make that part of the world which raises poultry. I have seen it stated by a man who is regarded as an authority that breed makes not one particle of difference in output; others think the breed is the whole thing. One man states that the whole fowl economy is set toward reproduction, and that it matters little in what state the individual bird may be, it will inevitably tend to reproduce, and will not suffer in this function because of disease in other parts. But a far more sensible view is that practically all other functions are dependent on the digestive processes. "The blood is the life" of every part, and on the digestive function depends the quality of the blood, very largely.

General average laying must needs depend to a great degree upon digestion; digestion must depend on proper feed and abundant supply of fresh air. Good health, established in a bird both through ancestry and in its own healthy life previous to maturity, must of necessity be better than a semblance of health that must be striven for because of unthrift in the previous generation or constitutional delicacy in the individual. The problem of raising phenomenal layers from the best in hand is difficult; the problem of greatly improving the average laying of the whole flock through rigid selection for thrift and through good care is one within the solving of every one who will take sufficient trouble.

An English writer, speaking of the fine averages made in the Australian competitions, emphasizes the fact that the greatest point made by the breeders of these Australian layers is strong constitution. He adds that the *lowest* pens in the year's laying competition have reached an average that would be considered good on most English farms.

General averages are the result of averaging other "averages," so that we need to know just what we are considering when we talk about "averages." For instance, at one of the best experiment stations, the average loss in stock for last year was 25 per cent in three-year-old hens, and above 20 per cent for all the hens kept. Most poultry raisers are willing to admit, I think, about 5 per cent of loss per year. Yet one of the big egg farms, carrying over two thousand layers, asserts that for an entire year not a single bird was lost!

With eggs, the matter is still more complicated, as there is a wider range of variation. From the general average of the farm hen, the country over, which our statistics assure us is in the neighborhood of six dozens a year, it is a long, uphill pull to the average, "from one hundred and forty to one hundred and fifty" which are ascribed to Mr. Prescott's birds. Mr. Prescott is a notable example of a man who makes circumstances bend to his adaptations, and his averages are the best I have known where such large numbers of stock are carried. When I was at his plant, he had about sixteen hundred layers, but a new wing stood ready to accommodate eleven hundred more. They were all Barred Rocks, and the flocks were the largest I have known. I doubt if another man in the country could get as good results with the same methods.

A step that strains our powers is that still upward to the claimed average of above two hundred for 50 per cent of the pullets raised, from a New England Wyandotte breeder. He also claims one hundred and ninety-seven as an average for the entire (large) flock. The one hundred and ninety-eight average of a flock of one hundred and thirty-five Barred Rocks in the hands of a Pennsylvania breeder, "in three hundred and sixty-five days from the time they reached laying maturity," is also notable. I figure that the average of the above "averages," without regard to the numbers performing, is a little above one hundred and sixty; but when we remember that there are hundred-millions of the low-average farm hens, and only a few hundreds, at best, of the heaviest layers, we may see how much nearer the six dozens than the two hundred mark the great general average must fall.

With little direct comment, I wish to refer to the most striking claims being made at the present time. It appears that, since several prominent editors have come out so strongly in asserting the mythical character of the two-hundred-egg hen — except as an occasional freak of nature — a swarm of claimants has arisen to say nay! nay! to these assertions.

The first whom I shall mention is L. F. Van Orsdale, a man who at least has the courage of his convictions. He reports one hen as laying two hundred and five eggs in her first year, one hundred and seventy-two in her second; at the end of which time the eggs were averaging thirty-one ounces per dozen, nearly a third more than the common market average desired. Another, a White Plymouth Rock, laid two hundred and twenty-eight eggs in a year. Mr. Van Orsdale has been trap-nesting and working for high averages for a series of years, and reports quite a number of hens as going above the two-hundred-egg mark.

Professor H. C. Pierce, Iowa State College, reports two two-hundred-and-ten-egg hens, one of which laid two hundred and four eggs in her third year, something quite unexpected.

The published record of U. R. Fishel's White Plymouth Rock hen "Irene" is two hundred and thirteen; the Cornell Station also reports a two-hundred-and-thirteen-egger, this record being made in two-hundred-and-sixty-one days, and beginning in January. The Maine Station reported, as covering several years' work, an aggregate of forty hens laying between two hundred and two hundred and fifty-one eggs within a year, —

these chiefly Barred Rocks; but one, laying two hundred and eight eggs, being a White Wyandotte.

A Wyandotte breeder of the Middle West, Ira C. Keller, of Ohio, reports a Silver Wyandotte with a record of two hundred and twenty-seven, a Golden Wyandotte credited with two hundred and thirty-six, and a White Wyandotte whose record reached two hundred and forty-three. F. Gage Cutler reports an average of above two hundred and forty-four with five White Wyandotte hens, the best of which gave two hundred and fifty-three eggs in eleven months.

J. W. Park, Altoona, is the Pennsylvania man who reported the average of one hundred and ninety-eight for one hundred and thirty-five hens, individuals among these being credited with from two hundred and twenty-five to two hundred and fifty-two each. Z. N. Allen, in a competition, reported twelve Barred Rocks as laying an average of two hundred and sixty-two eggs each, with the help of red pepper once a week. G. Redkey, Ohio, also in competition, reported eight White Rocks having an average of two hundred and eighty each; while W. S. Stevens, in the same competition, reported a pen of White Rocks as averaging two hundred and eighty-nine. I. K. Felch bought one of these birds, which he reported as having an average of two hundred and eighty-five eggs. The last great stride — to three hundred and thirty-four — is represented by a Barred

Rock reported by C. C. Loring, the New England breeder who has exploited the Buttercups in this country. In 1904 the Bureau of Animal Industry reported the "Sicilians" ("Sicilian Buttercups"), on the authority of Mr. Loring, as brought here in war times by a sea-captain (oft-repeated story), and that only one other importation could be found recorded. The birds are of medium size, and the eggs, according to this report, are said to be extra large and beautifully white. The Government writer adds: "The advocates of this breed believe it can, by proper selection in breeding, soon be made to produce three hundred eggs per year. It has already considerably exceeded the two-hundred-egg mark." This breed now claims a record of three hundred, which, if often duplicated, would doubtless make of it soon an American breed, at least as to distribution.

It has been said that, in a previous generation, a man who talked of two-hundred-egg hens would have been regarded as a fool. Also, that many a man is set down as a fool to-day, if he ventures to claim that the three-hundred-egg hen is a possibility. The writer concludes with the affirmation that, while there are to-day a few advocates of the three-hundred-egg hen, "they are generally regarded as a little off in their calculations."

There is little doubt, I think, that the average of the six-dozen-a-year poultry raisers look upon the men who report two-hundred-and-seventy-five and three-hundred-egg hens as actualities, as being Munchausens of purest descent. And it is certain that stories of the three-hundred-egg hens lead directly to such tales as that recently in print, of the big celebration in a certain foreign country because a phenomenal hen had just laid her thousandth egg. It was said — I do not know with how much truth — that our Government took the trouble to follow up this story, to find out how much of it was true.

Another golden nugget of the same mining appeared recently in New York. It was dated from a New Jersey town, and described the wonderful hen owned by Mrs. W. A., of that place, famous through having produced one hundred eggs a month. It was stated that this hen frequently laid seven eggs a day. The vivacious correspondent who reported the story took the precaution to kill the hen before the story went out; which was, perhaps, all the commentary needed!

A New England man once reported one of these abnormal layers, at a time when such reports made much more stir than they now do. He trap-nested assiduously, kept careful records, and made a study of his birds, in the interests of progress along every line. He was a keen business man, and his conclusion was that the two-hundred-egg hen would reproduce herself just about as often as the ninety-five-point fancy fowl, a flock of two-hundred-egg hens being about as easy to get as a flock of ninety-four-point hens. He contended that to force a fowl and to continue the process beyond her natural ability would result in the same deterioration that forcing a race-horse to do the work of a draught-horse would insure. "With hens, it does not always result in the death of the fowl, though I am sure the breeder is fortunate who gets his proof so quickly." Meaning that the quick death of such a bird would save the owner from injuring his flock, perhaps irremediably.

When we remember that the flock which sets out to lay three hundred eggs each in one year, has only two hundred and seventy-five days in which to make this record, exclusive of the three months usually allowed for the moult, it certainly looks dubious. And if any bird lay six days out of seven, as a usual thing, or about twenty-five eggs a month, say, she will be obliged to hold this pace steadily throughout every month in the year. From this showing, it looks as though "the masses" need not expect to raise three-hundred-egg hens until we find some means of omitting the moult. At present, the moulting period must define for us that "limit" of which the writer first quoted in this chapter speaks.

The "egg type" that has been so much written of has had many defenders, a few questioners; ten years ago there were few of the latter at any rate. Later, they arose in rather goodly numbers, and at one time it looked as though this favorite idea of a type of fowl which would show by her build that she was an extra-good layer — a type toward which all utility egg workers ought to aim — would fall to the ground. Now it is coming up stronger than ever, it seems. The oblong body, with equal breast and posterior weight, divided at the shanks, is urged to be the desired type, and it has been said that the pictures of the best layers would prove the contention. Before I saw this affirmation, I had been struck, in looking at the pictures of the leading claimants, to note how *very far* these pictures were from the described "egg type"! With perhaps one exception, as shown, they overbalance most decidedly at the breast; the "three-hundred-and-thirty-four-egg hen," especially, shows scarcely a third of her weight back of the shanks. The position in which she is photographed may tend a little toward this appearance, as one foot and the body are thrown forward; but even allowing this, the heavy anterior weight must be allowed to be much **exaggerated** over that of average fowls. We could surely wish it otherwise, for all poultry raisers would like to be assured of a definite egg-type toward which they could breed with confidence.

EGG FOODS: CAN THEY INSURE US A TWO-HUNDRED-AND-FIFTY-EGG HEN?

The Hen which appreciates Publicity — Station Work and Bulletins — Carbolic Acid and Cantharides — Formulas — Condiments will not do for the American Hen

It was a very shrewd writer who, having very little experience with the poultry business, as compared with those who have devoted their best years to it, produced a book filled chiefly with the experiences of others, which sold like the traditional hot cakes. Why? Just because it promised something hitherto impossible, at least to the majority — the two-hundred-egg hen. For the same reason, thousands are ruining their birds by feeding heavily with stimulants declared to be egg producers. Up to three hundred and sixty-five eggs a year (with one extra in leap year), appearances indicate that some promoter will in time go with his promises. Beyond that — well, time will show.

For a forced product, forcing foods, of course! The sale of them is incredibly large, and they go chiefly to town beginners and to farmers. Some time ago, I went into the salesroom of a Farmer's Exchange, in another state, and the most conspicuous ware, occupying large shelf space, was a certain poultry and stock food, "guaranteed" to do the impossible.

Printers' ink is responsible more than any other one thing for such a state of affairs. It is used freely in legitimate stories of experience, records of experiment, instructions for working, etc.; in advertising, in exploiting a thousand and one schemes for getting the cash of a more or less willing public. But Uncle Sam is the most voluminous editor in all our broad country, and his efforts lie in attics, rest on shelves, unused, mayhap bolster up the baby when she needs a higher chair; but at any rate, do not receive the study or the credence which they deserve.

The one great value of the Government and state bulletins to the farmer or the beginner in any line of work is that they are conservative and, in the main, consistent. It is charged by those who stand in the front line of instructors in poultry culture that the experimenters who get out these bulletins are doing primary work instead of studying up the knotty problems which bother advanced workers. This is often true, and I think we need to be thankful that it is true to quite a large extent; for it is primary teaching that is most needed by the rank and file, and it is conservative statement alone that will save thousands from putting money into poultry that they will never get out again.

Recently printers' ink gave to the public a statement

from a writer fairly known and favored by a few journals that *any one* who acts with intelligence and care can get *at least* two hundred eggs a year from each layer that he ought to keep, and that his own results have always been more than four dollars profit a year per hen, beyond the cost of the feed. This is supposed to be at average market price. I doubt if there is another in this country willing to father such a statement.

However, it was not this that stirred me up to this writing, but one or two other items in the public press. The first was from California, the place of large happenings, especially in poultry, and records that a hen of one of the newest breeds laid an egg every day from the first week in June till the following February, and between February and the middle of August weaned three broods of chicks and was again laying. The companion item concerned a hen directly descended from the flocks of a man who used printers' ink in advertising — a hen which very evidently knew what the publicity situation demanded. She had laid, as the story went, beginning at mid-winter to a day, two hundred and twelve eggs in two hundred and twelve consecutive days, and was still laying when the report was sent in. One can only guess that the story was not withheld till a better one could be told, because of the fear of the owner that a bigger tale would scarcely be believed by an intelligent public!

In any matter so far from the general experience of the thousands of people who raise poultry every year as the making of egg records of more than two hundred eggs a year, I do not believe that private records that seem phenomenal should be published. They can work only harm to the great majority of readers interested in poultry returns. And it certainly is good sense to take with a little salt any prodigious story issuing from a private source, particularly when the teller has an axe that might be sharpened by public credence. In other words, it is decidedly better for the average reader, and more particularly for the beginner, to study authoritative bulletins issued by men whose work is before the public, and who have no axes to grind. Some little allowance for human nature must always be made, even when considering state and Government work, but at least the chances for error are likely to be much lessened.

Some very excellent work has been done at several of the state Experiment Stations, and the results are available to all who care for them. It is true that one Station was called "the chief offender in the propagation of the two-hundred-egg fallacy," but most of them are careful and conservative. Their bulletins are safer guides than the flaming advertisements of those who have something to sell.

The following letter of inquiry may serve to show into what difficulties a too strong desire for winter eggs may lead the zealous:—

EGG FOODS

"Will you kindly advise me through the columns of your valuable paper in regard to my poultry? They were doing finely this spring, as we were getting forty to fifty eggs per day from sixty hens. We fed, all the winter and spring, bran, cornmeal, and potato mash, ground meat and bone, with some of the 'poultry food' which I saw advertised in another farm paper composed of cantharides, ginger, gentian, capsicum, Venetian red, sulphur, charcoal, and oilmeal. I gave whole corn at night; also oyster shells and grit with plenty of chaff to scratch in and a box of ashes for bathing. They have a large, light house, not overwarm in winter.

"The 'poultry food' did the business while it lasted, as my hens went from two or three eggs per day up to forty and fifty per day in two weeks' time, but since we stopped its use they nearly all seem to be ailing. One White Leghorn had swelled head all winter. About a month since, they began to dump and lie around and sleep, and we only got ten or twelve eggs a day. Just before this we stopped the mash and 'poultry food,' only feeding corn and oats nights, as they had full run of the farm. Several Leghorns and Minorcas died with bowel trouble. Some were crop bound, and the 'Rocks' nearly all began to swell in the head and throat and are not able to eat. I pour gruel, with a proprietary 'poultry food' down them; also put carbolic acid and belladonna in their drinking water and give a little chlorate of potash; also kerosene and turpentine dip for the head, but they do not seem to improve.

"I remember that about the time their heads began to swell I mixed some very salty chili sauce with their feed one morning. Could the trouble have come from that? My sitting hens and those shut up with broods were not fed the mash and are all right yet. What's the remedy for poison from salt? About one-fourth of the hens that were fed the salt are affected."

Reply: Here is trouble enough, and I am afraid some of it has been caused by a misconception of advice, or by a serious blunder of some one. Why any one should recommend cantharides as an ingredient of a poultry food is beyond me to solve. The only food ingredient in this so-called "food" is the linseed meal. The rest are drugs, and the cantharides is the apothecaries' "blister flies," so powerful that those who collect them are in the habit of using gloves and veils for protection, and of which the hundredth part of a single grain, when powdered and of full strength, is sufficient to cause blistering of the skin. It is even said to be dangerous to sit under the trees on which many of these flies are collected, and the volatile substance which they emit causes inflammation of the eyes and lids, irritation of throat and bronchia and often convulsive sneezing. Taken internally, they cause heat wherever they pass, and in large doses, inflammation and even death. It is as a blistering agent and in hair preparations that cantharides is oftenest recommended.

You will notice that your report of the difficulty with the hens, the trouble being chiefly with the head and throat, follows closely the line of what might be expected from the use of cantharides. Yet, as you say that this "food" was used all winter and spring, the question at once arises, "Why did not this trouble appear before; indeed, why did it appear only after the stuff was discontinued?" I can only conjecture that the dose was small, and that the effect was cumulative, having reached the fatal point just at the time you were ready to discontinue it; the resistant power of the birds being killed by it.

There is another point on which I am at a loss, and that is the quantity of bone meal and meat used. If this was heavy, and was suddenly discontinued, in this alone would perhaps be sufficient cause for drop in egg yield in case there were not insects to take its place.

Beyond all this, however, it looks as though these birds have genuine roup in one of its forms. The particular form which has the symptoms you mention is difficult to cure unless taken early, and is likely to be neglected as of little importance till it gets fully established, when it will be a slow fight to get rid of it, as it may run for several weeks, sometimes months. The chief reason of this, I apprehend, lies in the fact that the seat of the disease is the portion of the head over the roof of the mouth. Some natural discharge ducts being stopped by inflammation, the products of this inflammation are pushed out toward the eyes and around them, where they have not much outlet. Directions for treatment usually include lancing these abscesses, if we may call them such. But I speak from experience when I say that this is not ordinarily feasible. The skin is tough and the contents of the lumps hard, so that often the results of lancing are not satisfactory. The only successful treatment of which I know (beyond internal use of aconite, etc., and external applications of turpentine or kerosene) is through the nostrils and cleft of the mouth. Needless to say, this is not pleasant to the operator, and the birds object strenuously, as it is painful. A long-tubed machine oil-can, by means of which cleansing medicines can be injected into the nostrils and roof of the mouth, is the handiest tool. A feather will reach the roof of the mouth, but is likely to bend and slip aside, not reaching the point of difficulty. Perhaps in a majority of instances there will be, at later stages of the affection, canker in the cleft of the mouth. Sometimes this may be removed, but oftener it clings too closely to be removed without causing much pain and bleeding. A drop or two of muriatic tincture of iron, applied directly to the canker, will brown and shrivel it. Repeated applications twice a day will cause it to drop off and leave the mouth fresh and clean. Unless this canker is conquered, the bird will not improve. Canker is quite apt to appear first in the corners of the mouth, or on the tongue. Here it is easily handled, if one will take the trouble and be persistent. But let it reach far into the windpipe, and few cases recover. Sometimes it appears first in the windpipe, and unless the bird is very valuable, she might better be killed at once. Permanganate of potash and many of the roup cures are

EGG FOODS

good, if used as above recommended. When the canker is only in the front of the mouth, it may often be handled by using the roup cures or one of the two drugs I have mentioned in the drinking water. But up in the cleft it demands local applications in full strength, and these are always more quickly effective on canker than are the dilute liquids.

Applications of kerosene are excellent for swelled head, but need not be continued every day. In that case the head would soon be bald. It is imperative to separate the sick from the well, and to right, if possible, the ill conditions which brought on the attacks. Cooping in a shed coop, open away from the wind, is the surest road to recovery. Another eager correspondent wrote:—

"Can you give me a formula for the proper proportions of red pepper, charcoal, and oilmeal to make a good powder to feed for egg production? Is Venetian red of any value for eggs or is it only used during the moulting season?"

Reply: Inasmuch as the two great inquiries at this season concern getting plenty of eggs and curing roup, perhaps we cannot do better than to consider the ingredients commonly used in egg foods and roup cures, thus getting a foundation upon which to work out our own formulas. This is infinitely better than to give the number of ounces of this or that which may be used to make a certain "cure" or nostrum.

In the making of egg foods, a study of the subject will show both drugs and foods proper used as ingredients. The foods have food value, nothing more, except as they help in the better distribution of the drugs through the mash. The drugs are often stimulants, and charcoal is added to most preparations because it assists digestion. A formula which I used to some extent a number of years ago contained sulphur, ginger, fenugreek, red pepper, linseed meal, parched corn, and wheat. The parched grains were a substitute for charcoal. Linseed meal, or "oilmeal," as farmers are more apt to call it, perhaps, is always good *at linseed meal prices.*

When we come to the drugs, it is quite probable that the red pepper is the only one that is desirable. Fenugreek was formerly considered an essential; but its strong odor always told the secret, when it was present, and it has also lost much of its repute, so much so that the author of a book on "Popular Names of Plants," published in 1881, says that it is "now only used for giving false importance to horse medicine and damaged hay." I do not think there is much doubt that the red pepper is the important ingredient in all powders fed to stimulate egg production. A little charcoal is always good for hens in confinement, and a filler like linseed meal is always good and necessary, unless the pepper is placed directly in the hot water before mixing the mash.

A rule I have seen given is that mash should be seasoned with red pepper in the same proportions as would be needed for human food, — possibly a teaspoonful to twenty hens once a day. Please note that I do not say that this is a good thing to feed, from all points of view, but only that it will stimulate to immediate egg production in cold weather. I have, however, seen red pepper strongly recommended in a health volume as the great cure for liver derangements. If it deserved this recommendation, nothing could be better to use for fowls, as much of the treatment, especially when they are confined, tends to liver derangement. But physicians themselves acknowledge that they know very little as yet about this important organ, and as unprofessional treatment of animals is based primarily on our knowledge of human ailments, it follows that we really know very little about treating liver diseases in fowls.

Most of the drugs used in the treatment of head and throat diseases of fowls are of a disinfectant nature. These, with tonics, and, occasionally, emollients, nearly cover the ground. Chlorate of potash, permanganate of potash, peroxide of hydrogen, and tincture of iron are most commonly used, and all are excellent. Peroxide of hydrogen seems more painful to the birds than most treatments. There is nothing much better, all around, than tincture of iron. Used clear, it will cure canker in a very few applications, and without the necessity of taking off the spots before application. Ginger, pepper, or anything that keeps up internal heat or stimulation is good for internal use. Quinine is most excellent. Kerosene is a first-rate application, both externally and to inside surfaces of the mouth.

I presume nearly every one has failed with some one or more of these, but if so, it is chiefly because of inefficient work. The birds must be placed in a favorable situation, and treatment must be thorough and regular, and must continue for several successive days. Only on these terms will any roup medicine perform a cure. And if the disease is allowed to run into the severer forms, it may become incurable, or the bird may not be worth curing. The treatment most likely to be effective is the one which reaches mucous surfaces, and while it is common sense to take the disease in hand before individual treatment becomes imperative, it does sometimes become necessary to use disinfectant solutions with a syringe or oil can, through the nostrils and the cleft at roof of mouth. When necessary, tincture of iron, one-third, vaseline, two-thirds, may be used as a salve, both on swelled heads and on inner surfaces of mouth; but it should not be allowed to get into the eyes. With cases taken early this may be all that is needed, if quinine is given as a tonic for a few days. Tincture of iron is also used in the drinking water — the ideal way to administer medication to birds.

There is a Venetian red used as a dye, and this is the Venetian red usually meant, as far as I am able to judge. But there is also a red oxide of iron, known to some as Venetian red. This is sometimes used as a tonic, and also as a disinfectant of drinking vessels in hot weather. Personally, I have used only copperas, and that a good many years ago, before I learned to let such things alone. As a disinfectant, copperas has a good place, but I do not believe either it or its near of kin to be good to feed.

Stations in Massachusetts, Virginia, Connecticut, Canada, etc., have tested the effectiveness of condition powders in securing heavy egg production. Iron oxide, ground bone, salt, sulphur, and the other materials noted above, enter into them, in addition to common feeds. Feeding stimulants and fatteners showed, at one station, that, while they encouraged egg production in the small varieties, the Asiatic types were only made too fat, and soon laid soft-shelled eggs. Rocks and Wyandottes incline toward this type in many ways, and fall into the class for which stimulant drugs intended to increase egg production are prohibited. The oily meals, too, must be carefully used, when constructing feeding formulas for the larger varieties.

MOTHER AND CHICKS

The Friendly Sitter — Some Defects of Temperament — The Perfect Mother — Disquieting Colors in Chicks — Downy Harmonies

In the matter of mothering any kind of young poultry, I think the American hen can be proved to be as superior to all others as she is superior to most others in productive capacity and general usefulness. I am familiar by actual handling and breeding with about all the American varieties except the Javas and Dominiques, and I have never found much chance for fault finding in this respect, in any of these breeds. The type with much length of leg and of neck, and with the slabbiness that characterizes the large, rangy breeds, and the too actively fussy and nervous type meet on happy, middle ground in all the American breeds. Choice of variety within breeds is quite largely a matter of taste, although it may easily become far more than this when breeders introduce outside blood into a variety or strain, as is so often done.

Given a Rhode Island Red, a Wyandotte, a Buckeye, a Java, or a Plymouth Rock, the chief point of superiority as one attempts to prove his favorite in the lead must be along the line of extra hardiness, extra compactness or extra capacity for heavy laying. This means that the American nation, as a whole, is wedded to the American type rather than specifically to a single breed or variety. Within this type, undoubtedly, the future great American favorite will be found. At present they are nearly all great American favorites.

What points do we require in a bird to suit American notions? Medium size, compact proportions, good constitution. We want a bird of sufficient hardiness to hold her own in average conditions, an easy keeper, a generous layer. Every one of the American varieties, now eighteen in all, fills this demand fairly well. One thing more we ask, in spite of the million-counted incubator, offering infant machine-made lullabys, and that is that our satisfactory bird shall be a good mother. "Good," as applied to the hen mother, takes not so much account of her individual production record as of her disposition. Being of the desired medium size, she must also be quiet natured, faithful, friendly toward her keeper, and motherly. Above all, she must not be fussy, for the fussy hen breaks eggs and tramples chicks in an undue effort to convince the public that she is faithfully doing her duty. If we ask the fellow who likes some other variety better, how the Plymouth Rock hen ranks as a mother, he will say: "She's everlastingly wanting to sit, and she is selfish." Ask him about the Buckeye, and he will say: "You can't break her up any more than you could break up the original rock-bound-coast Plymouth Rock, and she lays too small clutches before wanting to sit." Ask him about the Rhode Island Red and the Java, and he will intimate that they are too heavy and sit too hard. Mention the Columbian Wyandotte, and he will hint that she hasn't quite sense enough to get things right every time. But when it comes to the White Wyandotte, he has to admit that if one likes the sitting breeds at all, it is hard to find fault with the White Wyandotte as a sitter and mother. That is, in the hands of a good and quiet handler. There are some fussy folk who could make any hen fussy. The White Wyandotte can be moved anywhere one wants to have the hatchery located. She will cuddle two or three dummy eggs with careful affection and fluff herself amply over the full nest of genuine goods when she is finally trusted with them. She is careful, not too heavy for most eggs, faithful, and friendly. After hatching, she is carefully solicitous that the young have sufficient hovering. If neglected, she will almost talk in her effort to have faulty conditions righted by you. The Buckeye, too, is an exceptionally good sitter and mother, and if the unfriendly critic above mentioned has managed to find a weak point in each breed, even among the favorite American sorts, his critical attitude is only a fair balance to the overzeal of their friends. And, really, is it not a remarkably good breed in which only one flaw can be found when considering it from any special point of view?

A few nights ago a man went in late to his supper, grumbling that he would never start an egg farm with an American breed. Well, why not? I have heard this same man say that no other bird could approach the Columbian Wyandotte as a year-round layer. Just this was the trouble: It was about the first of June, when the sitting fever takes the birds by wholesale, and he was "all in," taking sitters to jail after his day's work should have been ended. If a man is carrying eight hundred birds of the sitting varieties, he may find fifty or sixty sitters almost any night near midsummer. But nowhere is the "back to nature" call stronger than among poultrymen, and, if we discard incubators, as many are doing to a considerable extent, we must have sitters to hatch the chicks.

Ten years ago, when all the trend was toward the worship of incubators, I. K. Felch was many a time jeered at, and his teaching discounted. This was because he contended that the machine-made chick was almost invariably inferior to the nature-made article, and that he, personally, could detect every machine-made bird when it came into his hands for judging, at or near maturity. Now, some of the manufacturers themselves are admitting that the machine is not the perfect mother. No, the American hen is the perfect chick mother, as perfection goes at present writing, and the eighteen kinds of her are only eighteen variations of near-perfection along this line.

A large number of those who may become interested in purebred fowls suffer considerable agitation of mind concerning the color of the new-hatched broods. Except in the case of white varieties, the colors of the down are quite as likely to differ from those of the matured fowls as they are to correspond with them. For this reason, many who are hatching chicks from purebred flocks for the first time are apt to think they have been cheated by the breeder who sold them the eggs. Questions along this line always come in from beginners, in the hatching season.

In the case of white fowls, the chicks do usually follow the parent in color, being white or light yellow. But the White Rocks and White Wyandottes often show chicks of so dark a gray just on the surface that a novice would be sure to suspect impurity of blood. Yet those who have the most of these chicks assert that they make the whitest birds when matured.

I presume the black varieties are most prone to make the beginner uncomfortable, as white markings seem so out of place on a black chick. Yet Nature seems to have the opposite idea: she is extremely fond of furnishing white markings on the chicks of black breeds.

Under the body and on the wings such breeds carry this disquieting white plentifully, on the downy chicks. Usually the feathers come in black; partly, at first, then wholly, as they mature and take on their grown-up coat. But at times some white feathers appear to annoy and irritate the breeder, since they render his birds useless for show purposes by disqualifying them for exhibition. An old breeder will not, however, *always* condemn a bird as not fit for the breeding pen, because of a touch of white, because he knows that a feather bruised when green in the "pin" may show a white blotch from this cause. The difficulty of showing proof is the burden here. Many consider a white patch on wing the result of accident, if only one wing shows it. But if the fourth quill, for instance, on both wings shows white, it is considered to be a defect of the blood.

Some of the most interesting of the infant chicks are those which show harmonies of brown coloring. Among these are the Brown Leghorns, the Partridge Wyandottes, Partridge Cochins, such Games as show similar combinations of adult coloring, and the Silver Gray Dorkings. Most of these bear much resemblance to the partridges, or the tiny golden pheasants. They do not cause so much disquiet to the novice, because a mixture of similar colors characterizes the adult birds. The Partridge Wyandottes are not quite so dark, on the average, as the Brown Leghorns, as I have seen them, and they are not nearly so pretty. Yet one can hardly describe the difference. It consists mostly in a spot near the eyes, which gives the chicks a staring look. The head is larger, too, and the chick not so bright in its ways when very small. One would scarcely expect the Silver Gray Dorking to have chicks so much like the brown-red types, but when hatching some of these for a neighbor in the same machine with my Brown Leghorns, I have sometimes had to depend on the fifth toe of the Dorking to prove the breed. The broad, dark, rich brown stripe from the head the whole length of the back makes these chicks — I mean all of this color combination — very attractive. The colors make them very difficult to photograph well.

I have seen Barred Plymouth Rock downy chicks described as "dark, sooty brown on the backs." But I cannot convince myself that I have ever seen many Barred Rock chicks that would meet this description. I would call them more nearly black, or blue-black, with a lot more white than one would expect who was not familiar with them.

Birds of the Buff breeds usually give chicks creamy or yellowish, as would be expected. The Buckeyes, which are darker when mature, do not always show the chicks as dark as would be expected. Many of them come almost like the chicks from white breeds.

But, as the darker female standard now insisted on prevails, I think the chicks will average darker in down than they have previously done. The under-color of the adults is a very beautiful shade.

The Indian Runner ducklings are funny looking little blotchy things when hatched, showing much yellow, but spots here and there of dull gray-green. At their first feathering the sexes are much alike, and it is not till the adult plumage is assumed that the males can be readily selected.

What are called "foreign feathers" in adult fancy fowls are very exasperating, and usually they will be laid to faulty breeding stock; but every breeder of experience with Brown Leghorns will tell you that chicks which have not been properly fed with nutritious food or which have been below par because of lice, crowding, or other reason will develop many white quill feathers. The novice raiser of such chicks would almost invariably charge the parent stock with being culls. Such white feathers *might* come from cull stock but their presence is not *proof* of cull ancestry.

HANDLING THE CHICKS

General Care — Feeding — Late Hatches — Handicaps — Shelters — Spoiling the Chicks — Indigestion a Menace

WHAT a climate! In 1903, the night of the 6th of April, I lost five hundred eggs that were in an unheated room, through chilling; at least they were "lost" so far as their capacity for producing chicks was concerned. In 1905, on March 28, I put up temporary board shelters for the chicks to shield them from the burning heat. Such hot weather in March I never before saw, and only twice, so far as I can remember, even in April. Yet every year we do have extreme changes, and we try to raise young chickens during the usual frost and snow of March, and during the heavy showers and hot sun of June and July, and even August.

Because of this extreme variation in conditions, it is well-nigh impossible to give directions for the care of chicks that will hold good throughout the season; and those who have learned to raise them passably well at one season often find great discouragements at another season, because they do not realize the great need for different treatment. For nearly everything in the way of conditions which is good for them in cool weather is detrimental in hot weather and the reverse.

HANDLING THE CHICKS

In winter and early spring, under usual conditions, one of the things we have to make sure of is plenty of heat. We make every effort to give all the sunshine possible, and we shelter from wind to the best of our ability. Where chicks are with hens, shelter from wind is one of the critical points for those that come in March. Some people say that it is foolish, or almost useless, to try to raise chicks with the hens, in March. But if they are sheltered properly from the wind and kept out of the snow, few chicks do better than the March youngsters. In an angle of the buildings, protected on north and west, I can keep chicks in perfect comfort, a month or six weeks earlier than they would thrive in the open. And, indeed, there is very little time when they will thrive in open, unsheltered spots, because about as soon as they cease to need shelter from cold and wind they begin to need shelter from the sun, even in the average season. I had a good lesson on the value of a few boards on edge the past winter. In a cold frame of last year, but entirely without other shelter than the upright boards, I wintered some pansy plants. Less than three feet distant were other pansy plants, unsheltered, except for a little mulch. The plants in the frame came through almost without a marred leaf, while those outside were so blackened and rotted that I thought most of them dead. The roots were all right, but the tops of four out of five were destroyed, the exceptions showing green stems, but not much else. And protection from sharp winds means fully as much to the baby chicks as it does to the plant, if not more.

On the other hand, it is just as impossible to raise a good proportion of a brood if the coop stands unsheltered from intense heat as it is when unsheltered from cold. This is a point which is difficult for many people to realize. Summer chicks will do well if they can have shade and water in plenty, in addition to proper food, etc. Large coops are a great help when there is not much shade available. I find the three by six feet coops most excellent for a hen with a large brood. It looks like extravagance, I admit, to use such a big coop for just a hen and chicks, but I am not in the least afraid to assert that the chicks saved in a single clutch will often pay the cost of new lumber to build the coop, and this at market rates. (General figures are always based on market rates in my calculations, because fanciers' prices are exceptional and could not apply generally.)

If biting wind is sure death to small chicks at the beginning of the season, no less is the absence of motion in the air sure death to them in the crowded hot nights of midsummer. Those who are not in the habit of shutting their chicks in at night would learn much by going about at dusk and studying conditions among the chicks. Stagnant air, rising ammonia, crowding to suffocation,

HANDLING THE CHICKS

and other evil things might be discovered in such a night round. These are a good part of the cause of much mysterious lack of thrift and of actual loss during the heated term. During April and the first two weeks of May, in an average season and climate, any one can "raise" chicks, because the conditions are nearly all favorable. But it is the rest of the year that tests the caretaker.

Lack of meat is sure death to a very large proportion of chicks of the breeds that feather while the body is still very small. When insects are plentiful, and there is milk for curds to add to the feed, there is "good luck," possibly, but when these are lacking, meat must be supplied. Yet a large surplus of meat is as likely to cause death as is the lack of animal food.

Overfeeding is one of the causes of much of the mortality in tiny chicks; yet some are underfed. And it is well to consider that an animal may be underfed while having a large amount of food provided, because of the lack of variety or of proper proportion of muscle builders. And the fact that instances may be adduced of good chicks raised solely on corn does not affect the statement. When good chicks are raised "on corn alone," it simply means that only corn was supplied, and that the chicks were able to find the other things necessary to a good ration.

An early English authority on poultry says that one

great cause of loss during the first two weeks is the fact that food is not supplied continuously enough. Digestion being rapid and capacity small, the chick gets hungry before its feeder thinks of such a thing, and then gobbles too much when fed. This writer argues for food being kept before the smallest chicks all the time. I think this is the best utterance on this point that I have seen. But I would not make this ever-ready food of a kind that the youngsters are especially fond of, but rather a combination of two or three that they regard as second best. This will help to insure that they do not overeat. We feed as much as the chicks run after eagerly. The number of times of feeding would affect the quantity. It is a good plan to keep a dish of cracked oats where they can get it at will as some are slow eaters and need a chance to help themselves till satisfied. I place such a dish under a rack, where the smaller and weaker can satisfy themselves, without interference from the stronger chicks. One very successful raiser keeps meat before them all the time. This is the dry meat product.

Charcoal is an absorbent of gases, a purifier, etc. It is an undoubted help in cases of impaired digestion, and regularly in case of yarded fowls. But if chicks have plenty of range, I do not think it a necessity, though experienced poultry raisers and fanciers on town lots generally use it regularly. At two months, or less,

HANDLING THE CHICKS

the chicks begin to scorn the granulated foods as too small for any use.

In feeding chicks, especially if of a mixed lot as to size, and many running together, I am coming to the practice of making the mash pretty dry — that is, of using moisture enough to swell the ground stuff, but not enough so that it shall ball together much.

There are several ways of securing any desired consistency. Thick, sour milk will make the mash sticky under most circumstances, as most farmers know. Cooking the grain, or a part of it, has the same effect, as does also the addition of raw egg. And by the addition of any chosen quantity of fine middlings or "red dog" flour, the mash may be made of any desired stickiness. For years I have been in the habit of making all mashes sticky enough to ball together just enough to furnish fair mouthfuls to the birds to be fed with them. But in the cases noted above of many chicks, both small and large, running together, I think it is better to err on the crumbly side, if on either. Otherwise, the smaller chicks are trampled unmercifully, and also robbed. Very crumbly mash can be widely distributed quickly, and this is a necessity under the conditions mentioned. Several long troughs or boards should be covered as quickly as possible. Then, with feeding crates for the small fry to visit at will, they can be kept fairly thriving, even among an army of comparative giants, though it is decidedly better that large and small have separate runs and feeding places.

A good portion of the loss among very young chicks is due to musty feed. Probably this is more true with those who buy feed of dealers than it is on the farm. Not only does musty grain or mash material do more harm in warm than in cold weather, but it is far more likely to be common. Bran becomes musty in a very few days, when it is warm and in damp storage. Place beside these facts the additional one that hundreds of people can raise chicks early, who fail later in the season with the same method, and we have good reason for at least suspicion regarding the available foodstuffs.

If the chicks are developing well, there is no need to bother with any of the prepared mixtures. It is quite largely a question of quality of range. The grass may be tough and wiry and insects few. In that case meat is a necessity — at least with some breeds. With those which do not feather till nearly two months old, the necessity would be much less for supplying meat early in case the range furnished little.

I do not advise any farmer to supply meat unless he has to do so. And in case it seems a necessity and eggs are cheap, I would use raw eggs instead of meat, while the chicks are small. After two or three weeks they eat so much this would be extravagant, but I do not reckon anything extravagant for chicks under two weeks old, if I cannot get cheaper stuff on which they will do equally well. I consider the chief handicap to raising young chicks on the grains to be found on the average farm lies in the fact that these cannot easily be had in granulated form, and they are too coarse for the youngest chicks. Beyond this, it seems to me that farm chicks have every advantage over town raised chicks.

If without ice, it is scarcely practicable to use fresh meat with any regularity or freedom, but dry, ground meats are always to be had, and the difficulty lies in getting such as are sweet and wholesome.

Constant use of the granulated foods is urged by the makers, and the foods are supposed to be prepared to suit such use, but it is pretty well known that chicks gain flesh faster on soft feed than on hard grain, and bran is cheap and a good regulator of the bowels. For these reasons I use the ground stuff as much as I think safe, supplying the extra meat on account of using only two feeds of the granulated product containing meat.

As to grit, I am much opposed to paying two or three dollars a hundred for it in a prepared food, when it can be bought for one dollar. Hence, I always aim to supply it separately from the feed. Where gravel is plentiful chicks will get along without grit while small, but it is a safe thing to supply, and if they eat it you have the proof that it is needed. If not, you know they are all right, and need not furnish more.

There is a wide difference in practice and opinion among those who keep fowls as to the proper period for hatching the chicks for best results. Of course, breed makes much difference, as the very large breeds take a long period to reach maturity; but, leaving out of the question the Asiatics, of which comparatively few are raised in this country, we still find people who think April is already late, and others who make a practice of waiting till the latter part of May, or even June and July, to get out the greater number of their chicks. The large and rapidly increasing number of fanciers who exhibit has been chiefly responsible, probably, for the increasing desire on the part of the many to get chicks as early as possible. The incubator has made it possible to raise early chicks in larger numbers than could have been done before the advent of the machines, and the universal desire for winter eggs has been another strong factor.

One of our experiment stations has placed both those who contend for very early chicks, and those who insist that June stock is most satisfactory, in the wrong. The assertion is officially made that only in the natural season, about the latter part of April and earlier May on the average, is there probability of best results. I have watched these things very closely for a good many years. I once believed that only the early chicks were certain to do well and be profitable; but further experience has convinced me that the matter is almost wholly one of handling, and is in the breeder's own hands.

In case the late chicks can be made to thrive as well as earlier ones, there is a great deal to be said in favor of hatching right on through the season, or at least until one has a satisfactory number out. Indeed, from one point of view, it is decidedly better for the majority of farmers not to hatch till June, or even until August. The handler of fowls who cannot get winter eggs is not justified in hatching early chicks for layers, for they will cost him twice as much as late hatched ones before they begin to bring in anything.

And, admitting that late chicks can be made to thrive satisfactorily, there are only two reasons that I can see for feeding chickens from March until February — as so many do — before getting any returns, when those hatched in July and August will begin to lay at about the same period. One is that they have to put on their mature coat during cold weather and are therefore more subject to colds if not properly shielded from wind; the other is that when laying has been very heavy and the season very hot, the parent stock will not give as good eggs in midsummer as it will very early in the season. It is a question which each must decide, whether or not these objections are so great as to overbalance the real advantage of quicker growth, less cost to raise, etc. The matters of lice, too much heat, trampling, lack of water, etc., can all be overcome by careful handling, and there are no chicks that thrive like the midsummer chicks if circumstances are made to favor them.

The greatest handicap summer chicks have to meet is trampling by older ones. Another serious one is that water is not kept before them all the time. One little suspected evil is too strong sunshine on chicks just from the nest. Since I learned to keep the summer chicks confined to their coops for the first two days, with careful semi-shade after that, I do not have those inexplicable losses that used to be so common. Of course lice can deprive us of all the summer chicks if we let them, but we don't need to let them. The need for water all the time, and shade at will, cannot be too forcibly impressed on the boys and girls who are interested in the chickens.

I have seen a brood thrive perfectly for a month and go all to pieces through overfeeding for one or two meals, combined with lack of water at the same time. And no after pains could overcome the blunder. Overfeeding may not be fatal, if water is always at hand; otherwise, a dry water dish may mean a dead chick, perhaps a dead brood. There must be watchfulness sufficient to replace in time the shallow vessels which are necessary with baby chicks (unless fountains can be had) by others both deep enough and large enough to insure that the chicks shall not suffer through lack of water, if neglected for some hours. No busy farm dweller can be sure to be on hand at the minute every time. There will be many calls, many neglects that can hardly be avoided. That "contingent fund" which every organization cherishes so carefully is nowhere more needed than in the poultry yard. True, it consists, in this case, of extra feed and water, but it means money in the end. And it means comfort to many small orphans, machine-made, perhaps, but persistent in their demands, and very much alive to the situation. But, alas! not long "alive" to anything, if left to the mercy of circumstances, or the neglect of their foster parents.

On our place we have much pleasure each year in bringing up several families of what, for want of a better name, are known as "box babies," often contracted to "boxers." No patent-brooder chicks thrive like these, none show so few losses, none are so tame ever after. So that, except for the greater care which they require, no brooders need be had for chicks which come after the weather is reasonably warm — warm enough to supply all the heat needed during the day. A good sized soap box is the only brooder needed for twenty chicks, with a screen for the top. A warm but loosely woven piece of flannel is the only cover needed, and at five or six weeks of age the inmates of this primitive "brooder" will insist on no longer receiving the treatment of babies. They then go to coops at night, like others.

The very large number of people who are "too busy" to care for chicks after they are hatched should never attempt hatching summer chicks. Neither are they for those who have their quarters already full of older stock. But those who have a place for them by themselves and time to care for them decently need not be afraid to hatch all they want. I find one little bit of head work helps very largely to overcome trampling by larger chicks. When we must raise both in the same enclosure we place runs made of twelve-inch boards around the smaller ones till they can fly over them. Then, by simply locating the youngest chicks furthest from the gate where the feeder enters, they get along reasonably well. The young chicks are not with the main lot at all till nearly two weeks old. The great rush is near the gate at feeding time, and is over before the younger ones arrive on the scene. When they get nearly to the big bunch the feeder is already on the way across to the location of the younger ones, and they follow, so that they feed separately to a large extent.

HANDLING THE CHICKS

There are ill turns of the weather; diseases, like gapes; lacks unsuspected in range or feed provided, which may ruin many flocks of chicks, and we need not quarrel too much with the raisers who attribute their failure to untoward circumstances. But there are hundreds of runts and deformed or unthrifty specimens, and even whole broods ailing, where these results are almost wholly due to lack of care. It may even be that these chickens are well fed, and fairly well sheltered, as these things go where other work is put first.

A "fairly good shelter," in so far as protective value goes, is never a good shelter, if too small. Last summer I visited a place where poultry is made quite a specialty, and saw boxes in use for coops, beside which the average soap box would be a palace for size. The only time when a small box may safely serve as shelter for a hen with her brood is when there is a run attached for the hen. Otherwise she simply cannot get about without trampling the chicks, because they have no place of refuge; they must stay almost literally under her feet. Another point is that rain must beat into a small coop much worse than into one having fair depth. We aim to have coops of good size, then board up one end at the front as a rain shelter.

When midsummer heat draws on, many chicks are ruined for all their future by housing in coops that are too close and stuffy. Last week I saw a very nice arrangement on a large farm where hundreds of chicks are raised every year. It seems expensive, but if poultry brings any fair price, it will pay well. The summer coops were rather small, and slatted on three sides for free admission of air. A pair of these coops was placed under an open shed, which, later, will be the weaning coop for the two broods. While the chicks are very tiny this arrangement of two hens so near is not always safe, as they may kill each other's chicks; but after three weeks there is not so much danger of this. The plan might be worked by using runs at the first to keep the broods apart; but it is better for the chicks to range after the first week. They need not be doubled up till old enough to be reasonably safe.

The poultry raiser who has not provided himself with several of these sheds or weaning coops three by six feet or somewhere near that capacity has missed one of the greatest helps and conveniences. If for summer use only, convenience is best served by building them in sections, to be hooked together. In winter this has some disadvantage on account of cracks at the corners. But we manage to find handy uses for all of ours, winter and summer both.

Beware of corners! The incubator people have so nearly learned this that most hovers now take the circular form; but those who brood chicks with hens have mainly neglected this important point. The hen is nearly sure to stow herself and brood into the back corner of the coop. This works all right as long as she is with them; and if the coop is in "A" shape it is even better, since the chicks can slip back where the hen cannot step on them. Close watch will often show them doing just this little trick. It is when the hen leaves them that trouble swoops upon them. Just here is the very worst feature of chick handling on the farm, where people are usually too busy to see to little things or to think out little puzzles. The trouble is that in this climate we have many cool or cold nights following hot days. As soon as the sun goes down the featherless or partly feathered youngsters begin to whine with the cold. They seek the coops, often quite early, and spend a half hour, perhaps an hour in trampling one another, in the effort which each makes to be in the warmest place. If any one will take time to see the performance when the night has come on cold or damp, and see it through till the crowding is over, he will never again wonder that broods can hardly hold their own many times, or that weaklings die or dwindle. Only those that can crowd hardest have any chance to do their best, or ever will.

There are two ways to overcome this fairly well without providing heat. One is to board off the corners of the coop. Even heavy pasteboard will do, if it is to be temporary. A better way is to furnish board roosts as soon as the chicks will take them. They can keep nearly as warm by sitting close on a three inch or four inch board, laid flat, but a little up from the floor, and they cannot possibly crowd so badly. The general idea is that they will get crooked breastbones, which spoils them for the fancier, but I fail to see how a narrow flat board can be any worse than a wide flat board, so long as it is wide enough to support the whole breastbone, or nearly so.

Bunching chicks of odd sizes is pretty surely fatal to the smaller ones as far as normal development goes. If it takes all their strength and spunk to fight for place, how can they thrive? Last week a man told me, with every sign of satisfaction, that he had placed a second brood of chicks from the incubator in the same brooder with the previous lot, and that they were doing finely. Moreover, he was going to make them a run ten feet square; that was plenty big enough. Oh! the folly of it! After a little, he will wonder what he is feeding wrong, or why incubator chicks can't be depended on to thrive!

The town poultryman is, of course, the greatest sinner in the line of stinting quarters and green diet. His conditions make him so. But at times, in order to protect the crops, even farm chicks are confined too closely. It is safe to say that with average handling it is always unwise to confine growing chicks, except in enclosures large enough to furnish what is virtually free range; at least, an abundance of forage.

It is so easy to avoid these methods of spoiling chicks that one can hardly endure to see the havoc caused by just this little lack of care. When it comes to the matter of lice, the most potent cause of spoiled chicks, the fight is long and hard; longer and harder, as it becomes one of cure rather than of prevention. Prevention is easy; the only hard thing about it is to find time for it. "Cure" is wearing, wearying, discouraging, the season through. Three doses of insect powder, a week apart, on the sitters, go very far toward saving half the work of protecting the chicks. Lack of time goes far toward spoiling them in most cases.

"The pink of condition" has long been a favorite phrase with fanciers, to express the highest degree of apparent health, thrift, and "grooming." Except for the grooming (which often includes washing, blueing, fluffing, coloring legs, stimulating comb, manipulating crests and combs, etc.) it is within the reach of all of us on the farms to get and to keep our birds in this pink of condition. Or perhaps "the best condition" will be a sufficient phrase for us, leaving "the pink" to represent the grooming.

We are all familiar enough with the thought of the comb as an indicator of condition. When not too much animal food is given, the comb, being prominent and well colored, shows very quickly something of the state of health. *Exuberant health* is what we really need for birds that are to fill the leanness of our purses just after the holidays, and to obtain a flock of high, average standing along this line there is nothing else quite so effective as rigid selection, after we have done our full duty as to regular feed and care.

But the effect of meat upon the combs is quite deceiving at times. Sometimes heavy meat feeding will so stimulate the egg organs and the indicator of activity there, the comb, that the fowls appear to be in robust health when they are really being forced along this one line until affections of other organs are setting in, unnoticed.

Almost the whole exterior of the bird is an indicator of condition to those of experience and of careful observation. I have heard a man say: "That chick isn't right. I can't tell just how I know, but it doesn't look right, somehow." And again: "See what queer shape that chicken is; it is not going to live." This man was perfectly right in his conclusions, even though he could not tell why he formed them. The shape of the body in small chicks is often the first indication to a trained eye that there is coming trouble of the digestive organs. The posterior part of the body, say from the wings back, fails to develop, and the chick appears too broad for its length, this being many times accompanied by a little humping of the back. I don't think I ever knew such a chick to reach maturity. Usually these chicks die between ten days and a month old.

Something can be learned of the tendencies of the young before they grow sick enough to droop seriously. Broods that sit quietly in gossiping groups after eating, or stretch and sun themselves, are doing well. Those that do not sit down, but stand about aimlessly, are enduring some condition that means loss later on. It is the foresight or the quick judgment which sees these things early that enables one to be successful with the young stock.

The voice, too, at this time is a sure indicator, to the experienced ear, of the prospects for life. If it is a wretched whine, the chick is worth very little; but if it is strong and full, no matter if the chicks complain a great deal, it is more in the form of protest against what seems to them unfair treatment. An indignant, persistent cry usually means that the attendant is at fault in not providing feed or water liberally enough. And the chicks will cry for water quite as insistently as for feed.

How do you tell, the very day a chick is hatched, what its chances for life are? Many can do it; perhaps more cannot be so sure, and many would think this impossible. Not so. Partly by the looks of the fluff, partly by the fulness of the head about the eyes, partly by the shining of the eyes themselves, partly by the condition of the feet and legs, one who has handled little chickens much and has kept his eyes open will be able to tell from the first the chances of any particular brood, given intelligent care. Good care will always bring good chicks right along; but the best of care is helpless to do anything worth while for chicks that are sunken about the eyes and shrunken of feet and legs when first hatched. Skinny, shrivelled legs seem usually to point to something wrong with incubation. In such cases, the bill is quite likely to be shrunken, pale, or pinched, apparently, into greater than normal length. Roughened feathers and feathers coming in backward mean that something is not favorable to development.

What does it mean to you when you see a fowl stand with tail drooped sharply below the regular slant from the head when the body is well raised? What does a humped posture tell you? Do you recognize that there is severe irritation of the digestive organs to account for when you see a bird sitting or standing with its bill touching the ground? The sharp droop of the tail means, usually, some trouble with the reproductive organism. Roughness of feathers may mean lice, or it may mean unthrift from any cause whatever. The humped position may mean only that the fowl has a sour crop, and this may be overcome by holding it head down till the sour contents run out of the mouth, perhaps with a little manipulation. Then careful feeding for a day may bring it out all right. But the hump may mean much more serious things, and it belongs to almost every ill that attacks the birds.

Since good digestion is absolutely necessary to a fowl that is to be profitable, it behooves all poultry keepers who are still in the early stages of experience to learn to tell the condition of the birds, future as well as present, by the crop and the droppings. Normal droppings, unmodified by feeds that affect them specifically, are firm enough to keep shape, are brownish green, or thereabouts, in color, and capped by the white discharge from the kidneys. A normal crop is never full, and at the same time very soft, except when the bird has filled it with water. Such a soft, watery, or windy crop, if it become at all common with a bird, is an indication that you would better send her to the butcher about as soon as you can get her ready, if, indeed, this is still possible; for it means an abnormal digestive situation of some kind. It may be only a simple indigestion now, but a hen with chronic indigestion cannot long be a paying piece of property, for she is pretty sure to grow worse instead of better. The only exception to this rule which I know is in case of some birds which suffer unduly in reasonable confinement. Such birds, if placed on good grass range, may remain in good health a long time, when they might soon become worthless in confinement. Overfeeding soon means crops stretched and lax, and ruined birds. Too much bran or meat may make the digestive organs lax. Too much egg, if cooked, may cause constipation; if raw, it may still make the ration too rich, although, if fed but once a day, the danger is not so great. Bran alone (which has been fed in extreme cases) is not of good consistency, and has not sufficient variety of nutrition. Part middlings seems much better, both for hens and chicks, than all bran, and corn meal should be added.

MODERN WAYS OF HOUSING

Two Unusual Houses — A Square House Cheapest — A Hillside House — Cement-rubble Walls — Cloth-front Ventilation — Which Hen pays her Dues Best?

THERE are so many cheap books of plans for poultry-houses on sale at supply-houses and by the publishers, that I shall not trench on this field. But I wish to note some unusual plans, and to call attention to the principle of housing layers, and some modern methods of compassing it.

There may be two reasons for any particular form of house being unusual: one, that its disadvantages are too well known through its having been tried by so many different people; the other, that neither its disadvantages nor its advantages are well known, because the house itself is not at all well known or well tried. There are two of the latter class that perhaps merit the consideration of farm poultry keepers who have sufficient interest to build outright such houses as they may need.

The advantage on the average farm of the A chicken coop are so appealing that this coop is seen on practically every farm. Is there a possibility of extending these advantages to a house for a small flock of hens? The fact that one of the largest supply companies in this country is building such a house to sell, ready made, to its legion of customers, seems to indicate the makers' very strong belief that the A house will sell and satisfy equally well with the small A coop. To be sure, they have put a patent cap on the top, but air above can be secured in other and more simple ways.

The value of any house to any person becomes chiefly a matter of balancing its disadvantages and its advantages. If the latter are sufficiently great, considering the conditions under which it is to be used, the question is settled.

When one desires several small, disconnected houses, cheap, sharp-pitched, and easy to build, he may have them in a series of A houses, looking like enlarged coops, and capable of housing about twenty hens each, giving something over three square feet per fowl. If a square plank box, topless and bottomed with plank, be set on two plank runners and bolted fast, a firm foundation is assured. On this the A part may be spiked, overhanging a few inches. It may be shingled, covered with roofing paper or simply battened, and the front may be just as close as the owner desires. In those parts of the country where the colony house is the most popular, experience seems to have pointed toward shingling as most satisfactory. A man's first set of houses may be of various sorts, but his later ones are quite likely to be shingled. The varying price of lumber in different localities might have quite a bearing on this question. We have been using an asphalted felt, at about one and one-half cents a square foot, that is very satisfactory, and much cheaper than shingles. But one thing is true, I think, of all cheap roofings — their durability depends very largely on the care with which they are put on and the coatings on those that have to be coated. One year, in some cases, and at most two years, will be the "expectation of life" for a paper carelessly applied or left without proper coating.

Such a house as this is manifestly little more than a roosting room. A sheltered daytime run beneath may be provided by putting on a stout board at the back of the runners, in case the house faces away from the prevailing winds. Both this lower shelter and the house above have disadvantages, but many people would get along with them cheerfully in consideration of the many advantages.

The idea of the other house I have in mind is not at all new, yet it is seldom used. It has a good many advantages not to be secured, as well, in any of the popular models, and is cheaper and warmer. The alley in the poultry-house has been a great bone of contention, experience and foresight uniting in declaring its advantages, while the thrifty mind cries out against the almost waste of so much roof and floor-space. An ordinary alley is too narrow to be used for storage at all, and the expedient of letting the hens run under the alley or making the nests there does not seem to prove satisfactory.

Most people know that a square is the cheapest form to enclose. Hence, the nearer we can get our buildings to the form of a square, the less, proportionately, will be their cost. If the poultry-house is to be detached from the other outbuildings, convenience calls for storage room on the spot, and the house in which there is no long return journey after feeding will be the one that appeals most strongly to a feeder of considerable experience.

A square, if large, would give too much space with northern exposure in a house wanted for laying hens. An oblong, say fourteen to sixteen by thirty feet, is nearer the square than the usual long, narrow shed, yet gives good room for a storage and working place, without the large waste of the common form of the three-foot alley. A small room, it may be five feet by ten feet, cut out of the middle of one side, will provide an alley from which to feed, a storage room of some use, and a short route for the feeder, while leaving opportunity for all the sunshine desired, and pens lighted from three sides. If properly boarded, tight above the roosting platforms, it will give a good warm backing for the roosts, none of these coming against the cold, outside wall, as in the ordinary shed-long house without an alley. Unless the north wall is double boarded, I believe the placing of roosts against the north wall to be a mistake, notwithstanding the common custom.

A house in which work is to be done must have head room, and this form of house may have a double-pitched roof, unequal, seven to eight feet at the highest point. The short pitch will be sharp. The long one cannot be so, unless a roof window is used. This is not usually found desirable. It is scarcely possible to set them so that they will not leak, and not many people are found who pronounce them satisfactory.

Having for years studied every form of poultry-house plan known, I consider this, for a three-pen house, carrying from ninety to one hundred and twenty-five birds, about as satisfactory as any.

Here, where lumber is high, such a house could be built, covered with roofing felt, for something like seventy dollars, provided the work were done by a man at $1.50 a day. This is of hemlock, without floor. This seems a high price to those who are accustomed to house hens for little or nothing, in a shack that amounts to little or nothing, but the usual allowance for a fairly good poultry-house, as ordinarily built, without floor, has been for years about one dollar per hen when work is hired.

Fifty hens is so near the average number the American farmer likes to winter, that I think readers may be interested in a real American problem which was submitted by one of my correspondents: —

"I want a house that will accommodate fifty hens in winter. I am thinking of trying Rhode Island Reds and of excavating to a depth of two feet on a southerly slope, leaving a bank on the north side and the east and west ends, then walling to the surface with rock or concrete and using the same as a foundation, the scratching shed to be on the south side, and the house to be boarded up on the west end, from which comes most of our wind. I want to build so as to have one room in which to set hens, so that they may be away from the other hens. Will a long muslin window on the south be enough, or should there be glass also? The coldest weather is about thirteen degrees below zero, but we have mostly sunshiny days. In building the frame above ground, should I double board, or can I use tar paper on the inside? Would like to use tar paper to keep out mites, if it will not be injurious to the chickens.

"Lumber in rough is twenty-three dollars and twenty-five dollars a thousand. Lime and cement are each one dollar a hundred-weight. Would the stone or concrete become damp in cold weather, so as to cause sickness? I am new at the poultry trade, and any advice you could give me will be gratefully received.

"I want eggs in winter more than anything else. February 1, 1908, I had twenty-seven hens, and on February 1, 1909, I had sixteen hens, having sold down to that number, and giving them credit with just what chickens and eggs I had sold, they had netted me just $45.55. Of course, the scraps from the house and the milk with their feed were not charged against them, but we ate a good many chickens, and all the eggs we could use part of the time, and all we needed all the time, for which the hens received no credit, and I bought all their feed. I crossed Brown Leghorns with Plymouth Rocks, and they made fine layers and good rustlers and were of good size."

This was the answer: A hillside is always a tempting proposition to the builder of shelter houses, and, unless one has had experience, he is more than likely to build trouble for himself.

The climate and the character of the soil have some bearing on the matter. If you don't have much damp weather, you can risk something along this line; and on sandy soil there is not the danger that there is where the soil is heavy and holds moisture persistently. But our own main house stands on a gentle slope, and the soil is pure sand; yet in the spring thaws, and when there are heavy rains, we have to fight all the time to keep the water from seeping in. This house, however, has no stone foundation. The house built into a side hill, when built just right, has a drain entirely surrounding it, to cut off the water. Where this is not considered necessary, a drain is in the best practice laid under the walls and provided with an outlet for surplus water. When neither of these methods has been followed in building, the makeshift way is to bank up sharply behind the building, so as to provide what is practically a surface drain.

As to the floor-space, custom ranges from three square feet to ten square feet for each fowl housed. Three feet is crowding, unless the hens have open range. Conditions must always settle such questions. If you are to have a fine, open scratching shed and sunny weather most of the time, you will not need much light in the inner room, so far as the hens are concerned; but for ventilation and sunning, I think it preferable to have a window or a frame that opens, or a door, so that sunshine may reach the roosts and corners. Get sunshine in somehow, it does not much matter how, if the hens do not occupy the house in the daytime. Personally, I like one west window. It may not give quite so much warmth, but it gives some sunshine several hours longer. This for the apartment where the hens spend the hours of daylight.

I don't see that you need double boarding, though, of course, it makes it warmer. Hens sometimes interfere with paper that is down to the floor, of whatever character, according to my experience. You might board up fifteen inches, say; then use tar paper the rest of the way on the cold side. I never heard of its injuring the birds in any way. Why should it? The best-liked plan for securing ventilation and warmth at the same time is to build with a double pitch, put in a very loose overhead flooring just above a man's height, and fill in above loosely with straw. A door is arranged in the wall at the peak (sometimes an opening at each end). When this is open, the loose straw allows ventilation without drafts.

I will say once more that the question of cold and dampness in a fowl-house depends very largely upon the ventilation. Dampness, which you cannot shut out by any method of building, will come from the breath of the fowls and the droppings. A most practical method to obviate this difficulty is the open scratching shed. But in the roosting apartment it is, of course, the worst, and, no matter how you build, your closed apartment will be damp and cold unless sun and air get into it every day. Plan for this, and you will not have much trouble, I think, in sunny Colorado.

The Brown Leghorn always makes a good cross, I think. My own preference is for Rose Comb Brown Leghorns crossed with Wyandottes, or at least a rose-combed or a pea-combed fowl. But I never yet heard any one find fault with the birds that resulted from *any* Leghorn cross. If you get Rhode Island Reds, I think you can have some enjoyment from making a trial of the Brown Leghorn-Rhode Island Red cross. To use a few purebred birds for crossing, and the first-cross birds for the main flock is, to my mind, the most satisfactory method for the average person. Many people differ with me in this, but — they have that privilege.

Another American farmer, wishing to plan to throw two stones at one bird, — to vary the old saying a bit, — submitted his problem thus: —

"Please let me know if it would be well for me to build a hen-house with cement wall on the back. I have a nice lot of stone, the size of a hen's egg. I want to build a twelve-inch wall one hundred feet long. The front of the house will be of lumber. I would like to get rid of the stone. Please tell me how to build the cement wall."

Reply: To make a poultry-house which is, practically, a stone house, yet to have it bone dry, has been a problem to many a worker. I think it may be laid down as a first law that few poultry-houses can be made thoroughly dry unless there is a drain pretty well all around them. Some builders manage to get drainage by throwing up a knoll on which to build, or taking advantage of such a natural position, but natural knolls well drained *and also* properly sheltered are not so common as could be wished, and when natural conditions are not favorable, we must make them artificially favorable whenever possible.

Cement as a helper to the farmer and small householder and villager, indeed to practically all real-estate owners, is growing in use and favor by leaps and bounds. Just recently a farmer's wife of my acquaintance, having borne the last straw of inconvenience and dirt by reason of ground sloping toward the house, etc., demanded for the hundredth time (this time effectually) several bags of cement. She captured a passing mechanic, determinedly insisted that the "sometime" was now, and is now glorying in solid, broad porch floors of cement, solid walks from both doors to the street, and gates that don't sag. The last I saw of her she had an additional lot of cement in the wagon with herself and babies, and said that "Jim" had a box of the right size all ready for a "form" for a horse-block, which would be in use in a day or two. Cisterns, cesspools, water-lily ponds, even brooders for early spring chicks, hundreds of other handy and satisfying things, may all come out of the sand-pile and the cement bag, and I see them multiplying on the farms wherever I go, while it is reported that lumber is already lower in price because of the great popularity and usefulness of cement. Perhaps this is a trifle outside of my story, but I want you to have confidence in cement.

The house should be made with hollow walls, to secure dryness and prevent the passage of frost. The foundations should also be below the frost line. Excavate a trench fifteen to eighteen inches wide to a depth below the frost line, and fill with concrete composed of one part good Portland cement and three parts clean, sharp, coarse sand. Mix this thoroughly while dry, then wet and work up to the proper consistency to handle well and pour it over a layer of the stones in the bottom of the trench, ramming all down well. Be absolutely certain that every crevice is filled. Then add more stones and more concrete till the trench is filled. The stones should be clean, that the concrete may adhere thoroughly. For the benefit of others who may have no stones, I will say that the most common method of making concrete work where stones are not available is to mix with the cement and sand coal cinders or broken — that is, crushed — stone, the whole forming, on hardening, a solid block.

A firm of contractors has, within a few hundred feet of my residence, built recently upward of a dozen houses, and the foundations of every one are of concrete. They build a crib with heavy planks, well braced, just the size and height of the wall to be, make up their mixture of cement, sand, and cinders, and fill in the crib. Then the planks are left till the wall is hardened. Where stones are readily available, it might be better to use these instead of the cinders. But all should be rammed down well.

The building should have hollow walls. Most modern builders secure these by the use of hollow cement blocks, but this would preclude the use of the stone which our friend wishes to dispose of. As to the comparative cost, that would depend on circumstances, cost of labor, etc., which each must figure for himself. Of course, the wall could be built solid, and the inside be lined with boards nailed to furring strips; this would be easier to build, but would be more expensive, and I would prefer the hollow wall for sanitary reasons, if for no other. The wall may be built exactly plumb, or battered; that is, sloping in slightly from the base; but if built plumb, pains should be taken that it is exact. The foundation should also be a little wider than the walls.

The hollow in the wall may be continuous from end to end, in which case the two outside faces must be fastened together with rods passing through both walls; or piers may be constructed at intervals, forming solid places in the wall to tie the whole together. The latter is probably easier. In this case a "core" is made of such size as to leave the face walls about three inches thick; an air space of six inches is good. These cores are tapered slightly, so that they may be withdrawn easily when the work is finished. Sometimes they are made collapsible, which facilitates their removal. All being in place, the concrete is mixed as for the foundation. For an ordinary wall one part cement to four of sand answers very well. If forms are used the height of the wall, they may be left in place until the wall is well hardened; if narrow forms are used, which are raised as the work progresses, the concrete will usually be hardened sufficiently for this in twenty-four hours; but no pressure should be put on the wall under a week or two after it is completed. Not only the walls, but the roof, also, may be made of concrete. In this case, it should be reënforced with heavy netting or steel rods.

A slope of one foot in sixteen is sufficient. The floor, also, may be of the same material, and here would be an excellent opportunity to dispose of a goodly quantity of those stones.

To prevent such a wall cracking, it may be divided into sections by placing some thin material like a sheet of tarred paper crosswise the wall, but usually I prefer to risk the cracks. In case cracks do form, as the wall dries, they can be easily stopped with a little cement and sand. If such a building is well constructed, it ought to prove very satisfactory. I might say that I have seen a poultry-house constructed with a frame as though to be boarded up, but instead lathed and plastered, both inside and out, the plaster being made of the best cement, so that the walls were as hard as stone when thoroughly dry. Such a house can be painted any color desired, and is much more sightly than one covered with black tarred paper.

Perhaps the chief present-day interest in poultry-house construction is in connection with the cloth front. It is voiced in a modified form by one inquirer as follows: —

"I would like to ask C. S. Valentine's opinion of the Tolman fresh-air poultry-house or of any other up-to-date fresh-air house."

Reply: The semi-suspicion with which so many seem to regard the fresh-air poultry-house is based on a misconception. Such houses are *not*, usually, colder than the closed house, unless the latter is extra well built and particularly well ventilated. The ventilation has the last word to say about warmth, and no house filled with hens and kept tightly closed, without extremely good ventilation, can be kept warm. The air will have a damp, dead chill which goes to the bones, and this is the very thing to bring on colds and roup. Of course, there may be times, say, in the middle of the night in below zero weather, when the open house will show a lower temperature than a closed one, but the average temperature will usually, I feel quite sure, be found higher.

Mr. Tolman is by no means the first man who built an open-front poultry-house, but his house has some very good special features, and is the one that has, perhaps, been most talked about. It is well worthy of close study, since virtually all the reports we hear from it agree in the statement that both health and egg production are better in this style of house than in the closed houses previously used. The Tolman house proper is not a continuous house, and any one interested should put up a small one on trial before getting in too deep.

There have been numerous adverse criticisms of this house, and very many houses built which are modifications of it; indeed, Mr. Tolman has himself made quite a number of changes from the original form, and whereas he once stood for the wide-open house without reserve, he now admits that the colder sections of the country may require cloth curtains for severe weather, preferably before the roosts.

There are, however, some specific good points about this house which it is well for any poultryman to consider. One very little yet quite far-reaching one is that the floor of the roosting platform is placed above the level of the (low) front plate, so that it protects the birds from the air which enters at the open front. Another point is that the houses are set, not broadside, but end to the south, so that the roosts at the back end are well removed from the drafts of air that may enter. The open end is to the south, or a little east of south, and the window and door are opposite each other on the east and west. This lets the sun shine in pretty well all over the house at some time during the day, and it provides for central cross ventilation in the summer. In a very bleak and exposed place Mr. Tolman would use, instead of the usual inch-mesh netting front, a mesh of half-inch or even quarter-inch netting. This will keep out much more snow than one can easily realize, since there are so many wires for it to strike. The dimensions of the two recommended sizes are ten by sixteen feet and fourteen by twenty-four feet. The roof is double pitch, but with a long slope to the front, the peak coming just in front of the roosts, where he claims that the warm air banks and protects the birds. These are the distinctive points of the Tolman house, and I think the builder who believes he has a better one, which is any modification of this, has it yet to prove. The originator runs his houses entirely open, but says that "in very bleak and exposed places" it may be well to use a loose-woven muslin curtain before the roosts.

A goodly proportion of those who refer to the Tolman house have some so-called improvements to suggest, most of which can be shown by the originator of this house to be the reverse of "improvements." The modification most often suggested, or tried, is the curtain front to the building. Mr. Tolman says: "My advice is never to use the curtain, except in the most extreme cold weather, when the temperature is zero or below."

Back of Mr. Tolman, it is claimed, was the advice of a veterinarian whose help he had asked to get rid of a scourge of roup. He was urged to put his birds where they would get more fresh air, and was told that they would then tend to get well of themselves. Then this fresh-air house was evolved, with immediate results for the better. This type of house was the outcome of more than the ordinary amount of experience, and for this reason alone it is worth sufficient consideration to learn whether one can make it fit into his own conditions.

I have often made mention of especially good results in the use of the open shed with a small, elevated roosting coop at the rear. This is the nearest to the Tolman idea of anything I have seen, and it has been evolved independently, just through the demands of the daily work and experience. I think we had it in use before the Tolman house was ever heard of. Our own plan admits the use of the continuous house, whereas the Tolman plan is for colony houses.

One man wrote to the doctor above named that he had tried the fresh-air house and did not like it. But his own modification was quite a bit higher and only two-thirds as deep, and he did not have the long south slope of roof and the low front studs (five feet or less. Some build them as low as three and one-half feet front studs). This inquirer had missed all the good points, except the single one of the fresh air!

Professor Brooks, of Hatch Experiment Station, once tried a rather drastic experiment in fresh-air work. In a wire coop, covered, at one end only, with "a light roof of building paper," he placed a lot of fowls showing signs of roup. Over the sides of the roofed end was placed a protection of burlaps. "They recovered in a short time, and the hens began to lay in midwinter, almost literally while living in a snow bank." I don't think I should want to emulate this, but it shows what fresh air can do in the cure of roup.

A California man is advocating an "improved" fresh-air house which has one or two good points. That of a sloping floor of concrete, with a two and a half inch dip toward the front, could probably be adopted with profit by any one who uses houses open to the weather and without curtains. This is a two-story house — another attempt to attain the age-old desire of double space under one roof. The house is eight by sixteen feet on the ground. The lower portion is a scratching shed, four feet in the clear at the front. The upper is the roosting room, with trap-doors at the front for easy cleaning, and a floor rather ingenious in that it is level for two and one-half feet at the front, but thence it slopes upward to the extreme rear, which lets the droppings roll down and forward, and also gives six feet of head room at the rear of the scratching shed. He uses interchangeable curtain fronts, placing them on the scratching sheds during the day, or in front of the roosts at night in extreme weather only.

It is reported that a professor in the New York University, raising chickens as a hobby, has used the muslin front after a fashion of his own, tacking one thickness on each side of a two-by-four studding, thus securing double ventilation and air-space between. By the use of a self-registering thermometer he has been able to prove (not merely guess) that even in zero weather the temperature inside rarely went below freezing. Double-thick muslin would, of course, cut off much light, and he probably used no boarding at all. One point is not usually mentioned. I think I have never, indeed, seen any mention of the fact that a muslin front in a small house would apparently cut off more light than in a large one. There is one point sometimes made in favor of the cloth front that is not made quite fairly. For instance, the cost is compared with that of good lumber, without noting that the cloth must be replaced several times during what would be the life of the lumber.

A man in our own town who has tried muslin fronts had trouble at first because his coops were too warm at midday. The muslin was almost as bad as glass, when he had it permanently nailed to the front. But after putting it on frames which could be raised at will, he was much better pleased with its use. Where both cloth and glass are used, and there are doors besides, the cloth may be tacked permanently in place.

Where lumber is high and one has houses already built, I don't know a way that suits me better than to buy the ready-made "weaning coops" for use as roosting rooms. The kind we have are three by six feet, and about three feet high in front, floored, with a door at the centre of the side and a strip of wire netting along the top of the front. Burlaps or bagging can be dropped in front of this, if desired. We often find it good to use. Such a coop will carry twenty fowls, and they will lay well in January. It can be set on a roosting platform.

There is one point in a letter from another correspondent that has a bearing on the general discussion of cloth fronts in poultry-houses, and the differences of opinion with regard to them. "I have a warm roost," he says, "but the hens run on the barn floor." This is the ideal for poultry — a warm place at night, and a place to scratch and gather small grains throughout the day. It is the lack in making the night conditions sufficiently comfortable that causes much of the failure to obtain eggs, and in the cloth-front house this night comfort is imperative. The cloth front, of common unbleached muslin most often, was much recommended about a dozen years ago, as a movable closing to the front of a scratching shed, which shed was supplementary to a close house, small and warm, for roosting quarters. The muslin was intended for use only at nights and on stormy or very windy days, and was really only a storm shelter. There are a good many things yet to be said about winter poultry keeping, and it takes more courage to say some of them than most writers have, especially as they know that a storm of adverse criticism is likely to follow the saying. One of these things is this: In average sunny winter weather no other outdoor shelter is as warm as an open shed into which the sun pours; that is, having the wholesome warmth that is favorable to the health and well-being of stock of all kinds. When the sun shines, and the air is reasonably warm, — that is, in average winter weather, — an open shed is warmer than a cloth-front house with the front closed. It is not so stuffily warm as a double-walled house undoubtedly, because there is more circulation; and yet I am not sure that the thermometer would not register higher. It surely would if the sun shone on it, and the sun does shine on the hens in such a shed.

The average days, then, offer no difficulties; it is the extreme days, the dark days, the rainy and snowy days and the very windy days, when the wind is from the point facing the openings, in addition to the cold of late afternoon and early morning, that furnish the problem. Then it is a question whether the drop-curtain, cloth front, if entire, will be warm enough. Perhaps the larger number now compromise the matter, putting in one window, and a cloth-covered frame, like a large window, which may be closed on cold days, and opened whenever the outside temperature goes above that of the house. The general misconception with regard to the cloth front is that it must of necessity be too cold. But while air passes slowly and continuously through the cloth, it is an exchange process; heat trying to get out, cold trying to get in, and there is a continual battle going on, so that the passing off of heat is slow. With glass there is an active radiation of heat all the time, so that a glass window, unless double, is likely to waste more heat than the partial cloth front of the same size. Yet, when we come to the heart of the matter, there is no need whatever for a quarrel between these systems on the score of comparative cold, as either one of them could be shuttered, if thought desirable, except in the case of a full cloth front to a large shed, when this might not be feasible.

Even in a double-walled house cloth could be used over some of the openings in such a way as to be controlled in extreme weather, to the great betterment of conditions. Narrow slits near the roof might be covered with cloth, and thus serve as ventilators, but without the draught that is the bane of the poultry keeper who uses the usual style of ventilator.

A few days ago I saw this statement from a very practical man and editor: "Warm houses are still favored by most poultry keepers who try to get good results from their fowls, and roup is still very prevalent." A great many other poultrymen believe that the increase in roup in recent years is directly attributable to the hothouse conditions in which the birds were increasingly kept. This is supported in some degree by the disgusted remark so often heard on the farm, "I never had a case of roup on the place till I began to take special care of the hens and baby them." Or, it may be, "till I got fancy stock," in which latter case the stock gets the blame. The amount of it is that hothouse stock must have special conditions, and no break in the care; it cannot stand ups and downs, and its progeny may not be worth much.

The manager of a large poultry-farm says: "You can always tell a hen that has been comfortable all night. She gets off the roost in the morning feeling rested and cheerful, starts in to make trouble with her neighbors, kicks up the litter, and, later, steps up to the box-office and pays her dues." This is an expressive way of stating the facts which a close observer sees.

But there is a legitimate use of cloth in the poultry-house, over which there need not be so much difference of opinion. As long as men have kept fowls in separate pens, so long have they been scheming to cut down the expense of partitions. With males in two adjoining pens, the partitions must needs be opaque below, though I have seen inch-mesh netting used on both sides the dividing uprights. This makes all light available to all pens, but does not cut off draughts, and it is expensive. In one very large house I saw laths nailed close together to make partitions. Such muslin as is used for fronts would be cheaper than almost anything else, and it would allow the passage of some light. It would not be subject to decay through wet as in the case of the front, and the only real argument against it would be that it would attract lice. Still, it could be sprayed with ease, and would present no cracks. I should consider it decidedly superior to lath.

A large difference in results comes from variations in handling the fowls. If houses are low and overstocked, conditions will result which will be different from those following the housing of a smaller number in the same space. All who make a study of handling much laying stock learn after a while that the houses need to be opened to the sun and air freely whenever the weather will allow. They learn that the cloth-covered openings admit more light and air than they can get in any other way with so little disadvantage, and they also learn that a small window, in addition to the cloth, is a welcome addition on days when, though the sun may shine, the extreme cold or a wind from the front makes it undesirable to open the doors. The three kinds of opening give the best control of the situation under many varying conditions, and it is this control which settles the question of profit many times, if not nearly always.

There are many, of course (probably they are in the majority), who do not have the door in the front of each pen. But they try to secure the same result by framing the curtained opening so as to be opened when desired. They get approximately the same results, but not so much sunshine, light, and air as is possible with the three openings.

It is at this point that many of those who have condemned the cloth ventilation have made a mistake; they have used the cloth without either window or door, then condemned the system because it did not admit sunshine. I like the front doors very much, though they, too, can be criticised unfavorably from some points of view. (Indeed, what is there that cannot?) And there is this to be noted: the screen door is as much a necessity as the wooden door, unless the yards have unusually good shelter. On our own almost treeless, wind-swept, new place we cannot expose the birds without almost certain crops of cold and perhaps roup to follow. If there were sufficient shelter so that the birds were not subjected to biting winds, they could run outside in sunny weather with little detriment, unless the cold were severe. Unless they are free to go in and out at all times during the day, the secret of winter handling is to keep them in whenever the outside temperature is likely to be much below that to which they have become accustomed. In a sheltered location I should let my birds run free practically all the time from, say, 9.30 A.M. to 3.30 P.M., or a little earlier on the shortest days.

I have had one pen of birds located under a shed (partially boarded, but mainly wire-screened in front), and roosting in a weaning coop boarded over one-half the front, the rest being open. They disdained, in coldest weather, to snuggle in the boarded end, sitting every night opposite the screened portion. The proportion of eggs was better than in any other pen.

EXPENSIVE ACCIDENTS

Netting Traps — Cannibal Pigs — Paying the Price — Paying for Training

Possibly the American hen is a little less liable to some forms of accident than any others. This is because she is not so heavy as the Asiatics, nor so curious and "breachy" as the high-fliers. But every poultry-yard has its accidents.

A recent writer has put the whole matter of success or failure with poultry into a nutshell in the following sentence: "It is a business made up of small details and full of large possibilities, but most of us start out looking to the possibilities and overlooking the details, and the result is anything but what we desire."

If we wanted any confirmation as to the endless small details and the remarkable number of little things that one may do in the wrong way, we need only look at two or three of the books that have been written with the purpose of steering the uncertain in the right course. One is a compendium of five hundred questions asked by beginners, and their answers. A later one is similar, but claims to answer nine hundred and ninety-nine questions. A third has not counted the questions it answers, but is made up wholly as a reference book for quick information, and it has over two hundred and fifty large pages. This, in addition to the list, always growing, and numbering scores of books, dealing with the profit in poultry, or the handling, or the breeds, or incubation, and the like.

Just the chapter of accidents that cannot be prevented without minute attention to detail or a keen foresight for small difficulties is a long and varied one. A brief time ago a descriptive circular of a new wire netting drifted in. This was said to excel all other nettings in a dozen important points. I asked an old poultryman what he thought of it. He glanced carelessly at the picture, and turned away with an indifferent, "No good!" "Why do you think so?" "Diamond mesh, points up and down. Every fool chick that gets near it stands a chance of getting its head in the wide part of the mesh and shoving it down into the point till it chokes off its own breath." Though the new claimant seemed good in every point otherwise, every experienced poultryman knows that the old hand was right in his criticism.

I have seen a well-grown, thrifty cockerel killed instantly by getting a broad grass-blade across the end of the windpipe. I have seen chicks impaled on weed stubs, the end sticking out above the back of the bird.

I have seen scores of chicks caught by the toes in wire netting, heads hanging, and death sure if they were not released. I have seen them caught broadside between lapping wire nettings. I have seen both hens and chicks caught between coop slats which were nearer together at the bottom than at the top. They don't seem to have sense to shove their heads upward if they get caught. I have seen scores of chicks so scared by their first experience with a heavy shower that they seemed to lose their wits, and simply ran under something that looked like shelter, whether it gave any or not, where they were beaten to death by the drenching rain. I have seen big, strutting males put out of commission by a well-directed blow on the back of the neck by a weaker rival. I have seen chicks caught in all sorts of situations by something accidentally thrust through the punch mark in the web of the foot.

I have seen them drowned in drinking vessels, and choked to death by thirst, and paralyzed by the hot sun of a summer day. I have seen them taken with nervous spasms, perhaps the result of a scare, or possibly too much fresh meat, or who knows what? Last summer, one hot morning, I changed a bunch of motherless chicks, perhaps six weeks old, from one yard to the next. The new location had a big shed, under which they were placed, but left free to run out at will. Two hours later, while I was at work elsewhere, a quick and heavy shower came up. When I found the chicks, nearly twenty of them lay stretched, apparently lifeless, in the open, as near as they could get to their accustomed quarters just the other side of the fence. Had I not known that plenty of heat immediately applied would many times revive chicks apparently chilled to death, I should have lost them all, but quick application of heat revived all but three. Foresight will help to ward off accidents, and a knowledge of the chicks' habits of taking things will forefend them from many dangers, but some things seem beyond the power of the chick overseer to avoid.

In the matter of feeding, etc., there are many successful ways, yet there are scores of things one may do which will insure failure. I think the one thing which many fail most to realize is the insistent demand of fowls for abundant and continuous water-supply. Often what seems a failure in feeding is only a lack of constant supply of water. This will interfere with the digestion of full-fed chicks most seriously. The time to impress these things on our minds is before we begin the season. In the matter of eggs, too, this lack may be the one missing link. For hens cannot, surely, give a good supply of eggs without abundant water. Fortunate, indeed, are those who have pure running streams within ready reach of the fowls. Yet near-by water is not by any means always a thing for congratulation. People who do not yard their chicks or shut them in safely at night are full of tales of ill luck in having them attacked by water vermin, by skunks, by early crows and cats, and marauding dogs. In the town where I live a bunch of mischievous dogs ran amuck for a distance of more than a mile early one morning, and created havoc in poultry-yards all along the way.

Dogs are far from being the only ravagers. Last season an enthusiastic and almost inexperienced young farm housewife made up her mind to raise a goodly flock of ducks in order, chiefly, to provide feathers for pillows, etc. She hatched a fine lot, raised them almost without loss till very near maturity, and lost about twenty-five, after all her work was done, by marauding animals. It really pays better, if in any wise possible, to plan beforehand so that the stock is safe, even when the owner is not on watch. This does not always work, however, in the case of dogs, as they have been known to throw themselves upon wire netting with such force as to break it down.

Yesterday a heading, "A Business Man in Chickendom," caught my eye, and I read a bit to see what a business man would say. Concerning his own special breed, he affirmed: "Breeders will say that they furnish about all to be desired in a fowl — eggs, meat, or plumage." Well, "breeders" would say that of many another variety, but even yet the ideal bird has not materialized! Incidentally, the business man mentioned that he raised a few prize-winners. A little further along he stated that he raised them in A coops. The A coop has so many points to recommend it that those who see its defects still continue to use it more or less. But for a man who hopes to raise prize-winners to use an A coop is to throw away a good many of his chances for show birds. There is probably no other form of coop known that is so liable to turn out deformed chicks at the end of the season. This is because of the shape and the restricted space in the corners. This is just another of those details that are very likely to be overlooked. The "possibilities" of those presumptive prize-winners are so great that the small details necessary to their production are being overlooked.

At one time I received a letter from a man without health, without money, and without experience, asking my advice about keeping poultry. He stated that he did not even know one breed from another, and that he could find no one among his neighbors who was making anything from the poultry kept. Now, what advice dare one give such an inquirer, and what chance has he of success? The whole problem can be stated thus: It is not at all a question as to whether poultry does or does not pay, but as to how much it is going to cost to learn. The more I study the matter, the more I believe that this covers the whole field.

A man who is now having the best of success with poultry from the utility side only, said, in my hearing: "It cost me a lot of money to learn that hens won't stand dampness. I used to carry out dead hens by the half-dozen before I thought I could afford a drain around the house. Twenty dollars' worth from one house, in one winter, not mentioning the loss in possible eggs!" This twenty dollars represented what it cost him to learn that fowls will not endure dampness. But it did not need to have cost him anything to learn this fundamental fact. He simply was a doubter, who did not believe what he read.

My institute work brought me one night to talk with a woman who was keen to know about brooders. She was a bright, intelligent woman, who looked as though she would make things "go." But she told me that she had hatched five hundred chicks the last season, more than half of which were lost. At a moderate figure, those two hundred and fifty chicks had cost her, before they died, nearly twenty dollars. She had lost the possible profit of fifty dollars or more, had cared for twenty-five useless sitters, and fussed with two hundred and fifty chicks, for considerably less than nothing. Nearly seventy dollars and her work was the price she paid, in a single season, to learn how to raise chicks. Did it not cost her far too much to learn? Is it not poor policy to try such large numbers while still so ignorant? Why not learn with a smaller number? Whether she learned much or little, the cost was that seventy dollars. With a reasonable number to practise on, fully as much knowledge would have been gained, with a much smaller risk. It is even possible that she might have learned more with the less number, as there might have been fewer complications to throw one off the track of the real difficulty. Perhaps she could pay all this in one season as "tuition," and still make money. Prices would decide that. But even were this true, too much was thrown away needlessly.

It is by no means women alone who are making such errors. During the same trip I met a man whose poultry enterprise had been twice as large, and who had lost over five hundred chicks during the season. He had probably been too busy about other affairs to do the work himself, or else too hurried to observe closely, for while he had been attributing his loss to thieves, his shotes were eating chickens almost under his eyes and without suspicion. Is it not pretty safe to say that those who are quite ignorant of the work, and those who are too busy to attend to it, are almost sure to find it costing them too much to learn? The next thing is a conviction that there is no money in poultry.

Perhaps this matter will work itself out in the direction of public instruction in poultry culture. Experiment stations are taking up the work by correspondence, if not more thoroughly. Students have been seeking instruction in actual plants now in operation, only to find that those in the workaday world will not consent to be bothered with bunglers. But when it comes to a question of paying hard money out to a school of instruction, prospective students may ask themselves whether it will not be just as well to pay a pretty good price to learn in actual work in their own yards as to pay the same amount to a school, and then have much to learn afterward. The question is worth consideration. A study of it will go far toward reconciling the learner to paying for his knowledge, no matter where gained, and will make the general attitude more fair to poultry. If one would go into the work reasonably, possibly the school would not have many advantages over the other method of training. But in the school one may deal with large numbers of birds without taking any of the risk; and when such men as the president of the American Poultry Association could say that a half hour in one of the instruction classes made clear things that had been problems to him for fifty years, the fact is food for serious thought.

It has often been urged that those who would go into poultry work on a large scale should hire out to some successful working-plant. The advice is good, for those who have neither means nor practice in the work, if it is practicable. It is not practicable for a woman, for the only reason which could induce the owners of such a plant to take on green help would be the chance of getting rough work out of them. A woman would have to pay for her practice work, even if she could get in at all. I know of one large plant that refused to take a woman who almost begged for the chance to learn there.

The great point is to avoid, so far as possible, too heavy cost in learning. Often this can best be done by learning where some one else takes the risks; in other words, using another's brains till your own have sufficient training not to betray you into blunders which cost. But, if you must learn at your own risk, go slow; be fair, and, if you have to suffer losses, take them bravely, as a part of the necessary course.

What, then, shall be said to one who, like the inquirer mentioned at the outset, has much to learn, and is ill fitted to take risks? Shall we tell him to keep out, for his friends are right — there is no money in poultry? This would not be true, without qualification. There may be none for him, and I should advise him either that he keep out, or, if he insist on making the experiment, that he proceed thus: First, make up his mind how much he can afford to pay to learn; then invest not more than half of this at first, keeping the rest for contingent uses, and never spending, besides, more than the birds themselves furnish. This may not be the most profitable method of procedure, but it is the only safe method, where all the factors are either so uncertain or so certainly unfavorable to success.

COMMON-SENSE HANDLING OF COMMON DISEASES

Germs and Parasites — Preventive Foresight — Chicken-pox — "Distempers" — Samples of Difficulties Met Frequently

A LARGE proportion of the diseases of fowls, as well as those of human beings, are a result of the conditions in which they are compelled to exist. Some of these conditions are a product of the seasons as they pass. Summer diseases, like diarrhœa, sour stomach in some instances, and lack of thrift, with no apparent cause, are very likely to be due to the combined evils of excessive heat and lice, and lack of grit or of water. In such cases life becomes only a struggle to exist, with the American hen, as with all others.

The fact that nearly all germs and nearly all parasites increase more rapidly in hot weather would of itself render it certain that we should have more disease among our birds in summer. The fact that most outdoor workers are busier then adds to the probabilities of neglect and increase of disease through this cause, and the additional fact that most of the young are hatched at this period but emphasizes the probabilities in the same direction. It is perfectly true that most of the summer diseases can be overcome by the right kind of care, most of it preventive. But the fact remains that we are overwhelmed by the very diseases which we might prevent. And, worse than all, they soon reach a stage where they are virtually incurable. The white diarrhœas, however they may differ in character, the liver difficulties (including blackhead, which begins in the blind pouches), the state known as "going light," the tumors of the abdominal region,—these are mostly summer troubles, and they are largely preventable by sanitary precautions, and largely incurable when allowed to get a hold. They are also helped along by heat, lice, and lack of common comforts. It may seem strange to speak of tumors as summer diseases; but as most of the eggs are laid during the warm season, and as this is the time when a bird "off" in any way is easily gripped by disease, it comes about that most tumors become fatal in summer, no matter when they have dated their beginnings.

Nervous troubles among fowls, too, are much more frequent in summer than in winter. This is quite in accord with the nature of things, because the nervous system is in a more lax state in summer, and thus more easily subject to disturbances. Even the puzzling "limberneck" is now, by most authorities, ranked as a nervous difficulty. An outbreak may have a definite cause in the way of a heat-stroke or a bit of poisonous food, but the effect is apparent on the nervous system.

HANDLING OF COMMON DISEASES

We shall never be just to our fowls, nor successful in combating the wholesale diseases or the summer diseases until we realize that we must provide right conditions. We expect to provide for ourselves and for larger stock; we expect the hen to rustle for herself. She will do it without fail if we will provide favorable conditions, but even there we fail, and then complain that the hen is a nuisance. In all these troubles which I have mentioned it is just as well to recognize the fact that, if we have allowed the birds to come under the sway of these diseases by neglecting to provide the right sanitary conditions, we may as well save our strength and our money, as far as trying to cure them is concerned. In the busy rush of farm life in the summer season, very few of us can depend on ourselves to do all that may need to be done in the way of keeping a lookout, making right the first wrong things that show up, taking proper care of sick specimens, etc. Our only hope is in studying the habits and necessities of the fowls, and in making provision *in advance*, while we are not hurried, for such difficulties as might be expected to arise when we shall be in the midst of the rush. This is the only rational way of treating summer diseases.

I do not mean to say that there is no use of doctoring fowls. I am, however, quite willing to be quoted as saying that nearly all disease comes under the same head, in that it may be avoided by proper advance care. There are quite a number of diseases of domestic fowls that are curable, and which can be prevented from having permanent ill effects of the worse sort, even if we have been slack enough to let them get a hold.

Scaly leg is one affection easy to cure if one will give it the proper care, anointing with carbolic salve systematically till it is cured. Yet almost every poultry-yard shows cases of this affection, and few are cured, simply because it is slow of development, and the life of a hen is usually cut short through other causes before it reaches the fatal stage. This stage it may reach if neglected long enough.

Gapes is preventable by not raising the new chicks on affected ground. It is curable, or at least can be made tolerable for its victims, by the daily use of strong onion feeds, which are wholesome and not irritating or painful. Consequently, this is the most reasonable method of treatment. One year I dosed all my flocks with asafœtida, having read that it would take the place of onions. It could be used in the drinking water, and was cheap and easy. It worked well, apparently, for a while, but soon the chicks died as easily as the drug had been fed. I was young then. Now I prefer prevention to drugs, any day.

Some vent inflammations, even when they appear so bad as to seem necessarily fatal, yield readily to disinfecting ointments. The simplest thing for me has

HANDLING OF COMMON DISEASES

proved to be the use of zenoleum as a spray. A small spray-pump is kept charged with it, and any mange, external inflammation, fester, attack of mites, etc., is promptly treated to a spray of zenoleum. It takes but a fraction of time and is very effective, and is especially good when there is any ill-smelling discharge. It is pretty handy stuff to have on hand wherever live stock of any kind is kept. It is recommended also for scaly leg in fowls.

Early colds may be routed by the use of quinine or aconite in doses somewhat less than one would give a young person, and soft, swelled head can be routed by the use of kerosene applications. Sour stomach may need only a little soda, just as a child with nausea might take. Some of the sickest-looking chickens I have ever seen have recovered miraculously on being dosed with a bit of soda. But it is quite wise to keep an eye on a bird that has acquired the habit of indigestion, for it is rather sure to develop eventually into a worthless specimen, so far as profits are concerned, if no worse.

Chicken-pox, though it is often caused by filth and dampness, and though it is contagious, may be usually overcome by the use of ointments containing disinfectants.

Possibly there is no common disease of poultry which is as little known among Northern and Eastern farmers, or one which causes as much puzzling and alarm, as chicken-pox. After it has run to its height, it is both disgusting and dangerous looking, and it may or may not be as dangerous as it looks. For some reason not fully understood, it seems a capricious disease. Poultry may be raised on a farm for a long term of years without a sign of any such ailment, when all at once, without introduction of any outside stock, or other reason easily apparent, it will break out and run through the flock, sometimes virulently.

There are, however, certain seasons and certain conditions which chicken-pox seems so to prefer that those familiar with the disease usually attribute its outbreaks to these conditions. It usually appears in the fall of the year, and often when the young birds are slightly less than normally vigorous, through the ordeal of feathering or the hardships of the advancing cold. Yet the Gulf Coast of America and the southern countries of Europe have far more frequent and fatal attacks of it than those of cooler latitudes. At the North, poultry raisers are more frequently entirely unfamiliar with it; but if they are acquainted with it, they are often very indifferent to its presence in the flocks. I once went through the houses of a large poultry raiser in Massachusetts, where birds affected with this unsightly, infectious, and often fatal disease were running freely in the pens with many others not affected. Yet the owner was fully aware of the character of the affection, and seemed to have little fear of its spreading.

HANDLING OF COMMON DISEASES

Fowls which die from this disease at the North are very apt to have contracted with it some form of cankerous roup. Often they become blind in one eye, or both eyes, by their closure. They can eat little, if at all, and weakness supervenes, followed by almost certain death.

At one time this disease was considered to be simply a form of varioloid. Those who have seen the latter, or smallpox, or who remember the ugly scabbing connected with vaccination, should have little trouble in recognizing the chicken-pox in fowls. The yellow scabs, often foul when discharging, are generally found at first about the head. But they sometimes attack other portions of the body, especially the under side of wings, and it may be the inner surface of the legs. A child would probably call the nodules on comb, wattles, etc., "warts," at first. Later, after they break and discharge, they become confluent scabs. Pigeons and young chickens are particularly susceptible. It is said to attack a large proportion of these at the South, causing much damage and many deaths unless early and vigorously treated. A peculiarity is that these raised scabs often appear on the horny bill itself.

Any one who is not sufficiently afraid of it to be induced thereby to give the disease proper treatment, should familiarize himself with it under another of its names, viz. a "blastomycetes fungus." Perhaps this will scare him into doing his utmost duty to his birds!

Dampness and filth are believed to be at least contributory, if not producing, causes of this filthy disease. The germs need moisture in which to grow. Looking at it from all points, common sense insists that sanitary treatment is especially indicated. Disinfective ointments are recommended, what is handiest being the best, because it can be most quickly used. A good cleaning up, a rigid quarantine of the sick, and a close watch for new cases will often stamp it out very quickly. Iodine is a good application for the sores, but harsh. Probably the majority of those familiar with the affection use carbolated vaseline, as carbolic acid and vaseline are almost always at hand, and can be combined to form this ointment. One part carbolic acid to twenty parts of soft soap is recommended. In this case, warning is given not to cover a very large area at once, as the absorption of the acid might cause poisoning. Carbolic acid solution and lime washes are effective for disinfecting houses and runs. With due regard to their poisonous character, almost all the washes used for roup may prove effective with chicken-pox, because both are treated from the standpoint of disinfection. At the North many find simple vaseline applications to the sores to be effective. Perhaps this is on the principle which physicians now follow in treatment of many eruptive diseases, such as scarlet fever, that oily applications prevent the scabs from scaling off into the air as dry, fine particles, which contaminate wherever they go. Drinking and feed vessels should be scalded well each day while the attack persists.

There are a good many inquiries about "distempers," so called.

The word "distemper" really means disease. It is applied chiefly to animal diseases or any morbid state of the animal body. But as I understand the use of the word among the people at large, it carries the idea of some general, supposedly infectious or epidemic, trouble. In short, the word may have almost any meaning that includes disease, and I have used it chiefly to avoid using the word "roup." Roup proper, to my mind, means a virulent form of cold, which is also infectious, and the word is often loosely used when a simple cold is the only difficulty.

People wonder what can be the cause of so much cold and disease of the head, throat, and bronchial organs in poultry. There is often a predisposing cause within the bird which is not much recognized. I refer to the growth moult of young stock and the annual moult of the older birds. It is the simplest thing in the world,— in short, the bird loses so much plumage that it suffers from the weather; it cannot keep warm. Almost twenty years ago Mr. Felch called attention to a "distemper" that was sure to attack young stock at four or five months old, according to the breed, when changing feathers. This, he stated, was as sure to come as measles in children, and needed only a little bromide of potassium in the drinking water for a fortnight every other day to cure the difficulty "and ward off the roup."

If chicks reach the age when they assume adult feathers during warm weather, they will probably not be noticeably sick. But the great proportion of the young stock of the country is hatched late, and comes to this moult at the most trying season of the year, when the cold fall winds begin to blow. The later part of the season is pretty sure to be the season of neglect, as the poultry raiser feels that the chicks are at last big enough to care for themselves, and off his hands.

Every one knows that the way to consumption and pneumonia is opened for mankind by a neglected cold, "not worth mentioning," perhaps, at the first. Admitting that poultry are especially subject to diseases of the head and lungs, is it not to be expected that a little neglect, the conditions remaining the same, will soon cause a simple cold to become a fatal bronchial disease?

These colds, or distempers, take myriad forms, which pass from simple to dangerous so insensibly that no one can tell where the line is. There is just a frothing at the eye and a whitened tongue; there is a swelling of the face about the eye on one side or on both; there is stoppage of the nostrils with gummy mucus; there is rattling stoppage in the bronchial tubes; there is the dreaded canker, usually considered the worst form of roup.

The almost universal failure in the treatment of these forms of throat and head distemper is due to just one thing, viz., the birds remain in the ill conditions during treatment. Instead of being removed without delay to a house where dampness and wind cannot reach them, and which is reasonably warm and well ventilated, they are left under the conditions that brought on the affection. This is the one great reason for the failure of the patent roup cures and the never-fails given the owners by their next of kin and best friends. Nearly every one of these is good, provided the fowl is given a chance for its life by favoring it as to conditions. It is hardly judicious to place it in much artificial heat, as this tends to make it sensitive, and there will be trouble when it must return to the cold house.

Next is the question of medicines. One of the best poultrymen of the country has made the statement that about all we can afford to do is to see that fowl patients are removed to dry and airy quarters, unless they are unusually valuable (as fancy or show birds), and that "doctoring poultry with infallible cures for roup and other numerous complaints is time and money wasted." This in the face of the fact that the better poultry papers carry departments for veterinary advice handled by genuine medical practitioners.

The first question I ask regarding any medicine recommended is whether or not it can be used in the drinking water; if not, I look for something better. But canker will seldom yield without direct applications to the morbid growth. Moreover, I have never yet seen any medicine that will kill the canker that will not also hurt the birds' mouths. They resist its application, and one must be cruel to be kind. Kerosene is a good application for swelled head, but I prefer carbolized vaseline, with enough tincture of iron well stirred in to color it a little. This must not be allowed to enter the eye socket. Dr. Salmon, the Government specialist, who has written a book on poultry diseases, recommends that bad collections of pus about the eyes be opened with a sharp knife, the opening being washed out with disinfectant, or the wounds filled with disinfectant powder.

Recently I heard a man say that the locality near Jersey City was the worst for roup of any in his acquaintance. If there was ground for his opinion, it probably lay in the excess of strong winds which sweep down from the heights, and often in from the coast. On every poultry-farm where roup is a frequent visitor one should write in very large letters over the doors of the houses, "Look out for draughts."

A promising new poultry-house was built of unmatched hemlock and covered, both sides and roof, with

HANDLING OF COMMON DISEASES

tarred felt. The builder took great pains to nail the roof boards down tightly, and flattered himself that for once his birds were well housed. But in ten days all the young stock developed distemper. It was found that the roof boards had already warped away from the siding, and, as the roof was built to overhang, the covering paper could not easily follow up over the cracks between roof and side. The roof boards had also shrunk, so that a finger might be laid between them in places. Here were the precious pullets sleeping with a separate draught playing over them from every crack. The plate crack was then filled with tar, and every roof crack filled with tissue paper calked in — and still it leaked wind. Finally, to overcome the defect, the house was lined also with tarred felt above the roosts. Such is the care needed to make sure of keeping out the draughts.

One who has been having a few cases of swelled head in a flock of three hundred birds said to me: "We used to get lots of eggs, and we never had a sick bird; the hens roosted in the trees, and we never took any thought of them except to throw out a little corn occasionally." "Then, if you really believe that to be the better way, I advise you to get back to that way of handling the fowls as soon as possible," was my reply. The mild grumbler laughed: "I'm not sure that I do really believe it," came the answer. Many another has occupied the same critical attitude toward modern poultry keeping and housing; yet I do not find those who are willing to go back to the "survival of the fittest" plan, notwithstanding it has some things to recommend it. There are few methods of which something good may not be said; our problem in the twentieth century is to find the plan of which *only* good may be said. But we have not arrived.

A few of the letters which have come to me, with the replies, may serve not only to show the helplessness of many poultry keepers in the presence of disease, but may give an idea of the commoner difficulties, and voice a warning.

"I am in trouble and want a little help. Last fall I bought ten pullets and a rooster of the Asiatic breed, with feathers all down to their feet. The man I got them of fed all corn and as much as they would eat, and they were as fat as butter. He told me to do the same. January 1 my rooster dropped dead in his tracks, weighing fifteen pounds — apoplexy, I suppose, as we have no lice, no mice, no rats, no minks; nothing of that kind. The next I noticed my chicks commenced to call 'pip' when I fed mashed feed. Then some commenced to wheeze and rattle in their throats, so I sent for a roup cure and I took fat meat and rolled it in black pepper and put that down my affected hens and kept them isolated for a few days, and they got better. I got buckwheat, oats, and cracked corn and mixed it and fed that as a scratch food, and I boiled potatoes, apples, cabbage, and meat scraps (pork), and mixed it with chop and cut alfalfa. I also got a bone-cutter and cut up bone, and fed that.

I give them raw cabbage, raw apples, crystal grit and oyster-shells; also a patent poultry-food. I keep the coop clean and have a large dust-bath for them; and I got twenty-six eggs in December, eighty-four in January, one hundred and thirty-two in February; but now I am coming to the trouble that I can't overcome. Every now and then one seems to lose power in the legs for a day or so, and then come all right again. Some — most all — have loose bowels, so that they are soiled all the time. What can I do to stop that?"

Reply: The conditions here noted form a clear object-lesson, showing the truth of a statement which I have made more than once, that one must have some experience to found judgment on before he can be sure of exercising good judgment. The trouble with these hens comes chiefly, if not wholly, from lack of good judgment in feeding.

When you found out that the corn was too fattening, you were on the way to better handling, but the potatoes, green bone, the prepared food, pork scrap, etc., have been a snare to you, for some of these are as fattening as corn, and some of them are very forcing. You have simply overdone the whole matter. It is no wonder the rooster died, for, strange as it may seem at the first, it is more difficult to feed a cock than it is his mates. The hens always have an outlet for their surplus in egg production, but in the off season the males simply store up fat or tend to liver troubles, and often impotency, if overfed. The word "overfeeding" does not always mean feeding too much; it may mean feeding too much fat for the rest of the feed, or too much of the forcing elements. You had fat hens to begin with, and corn as part of the ration. Then you added potatoes — another fattener — and pork scraps, which were "worse and worse and more of it," as we used to say when we were children. Buckwheat is a good grain to feed with corn, on account of the fibre and the muscle makers in it. A little green bone added to your grain ration and the alfalfa and cabbage, etc., would have made an ideal ration, but when you added pepper and pork scraps, and the prepared food, or any other stimulating feed, you were running the risk of losing your hens by what might rather be called overdosing than overfeeding.

The simple feeds you have used are all first-rate, only that they were so combined as to be both too fattening and too stimulating. The grit, alfalfa, and green feed have probably saved you the hens. One has to be a little more careful in feeding Asiatics, especially with fattening things, since their tendency is to lay on fat, and they are slow in temperament, inclined to laziness. It is quite possible that the touch of roup was also due to the overfat state of the birds, for excess of fat diminishes resisting power and also tends to difficulties of breathing. I do not think Asiatics are much subject to roup when kept in robust condition. All who aim to succeed with hens need to learn to handle the birds often enough to know at least the average condition of the flock. Hens need to be in good flesh and fairly fat always. The smaller breeds will grow thin during the heavy laying season unless some effort is made to keep them in proper condition. And, with fair treatment, it is only in the interim between finishing the moult and beginning to lay that there is serious danger of the females getting too fat. Neither will they lay till they are fat enough. It is, then, plainly a matter which requires good judgment to keep fowls fat enough without getting them too fat at certain seasons. The Asiatics require a goodly amount of average feed, and to this one must add fat makers or egg makers as the season and the condition of the birds indicate. Nor will it do to examine one bird and decide from this, as some birds take on fat more easily than others. Good digestion and good appetite are the key to the matter, and these are always to be watched, as affording the ground for judgment as to needed feed.

The diarrhœa is plainly caused by the feeding. It may be that you have used too much green bone; from half an ounce to one ounce a hen is the usual allowance. This is generally given only three times a week. Potatoes and green food have a tendency to cause looseness, and fats are given to the human family as laxative medicine, oftentimes. If you want a feeding basis that will often be of help, consider how the things you use, or their equivalents, would affect human beings. These will not always apply, but will apply often enough to be a real help. Don't be one bit discouraged; you are on the right track; it is always more difficult to handle a few hens, as to feed, than a larger number. But it can always be learned, and you are getting your experience as you go along.

You can cut out the mash for two weeks, and soon get rid of the diarrhœa, probably. It may affect the egg yield a little. Drop the patent feed, which is highly stimulating, and also the fat meat scraps. The scraps would do to feed lean Leghorns, but will not do for fat Asiatics. Too much green bone causes diarrhœa, and you need to watch carefully that you do not use too much. As you have had a bone-cutter and so few hens, you would be likely to give too much, or else keep it stored till it was tainted. I don't say you have done this, but it would be a tendency of human nature; and I speak of it as a warning to you and to others who have only a few birds. Those who have large numbers have to scrabble so hard to pay their feed bills in winter that they are tempted to scrimp rather than to be overliberal.

If you have now a cock with the hens, it is quite likely that the temporary inability to move about is

HANDLING OF COMMON DISEASES

due to his attentions. Or it might be difficulty in laying, or possibly rheumatism. The first is the more common cause of temporary difficulty like this at this season of the year. I have known a hen to be permanently lamed in this way, but the difficulty usually lasts only a day or so, often only an hour. Rheumatism is more apt to continue for quite a long period.

"We have some White Leghorn hens that have been laying since January, and lately four have had trouble discharging their eggs, as I have found them with the egg out of the body, but not out of the passage. These hens or pullets were hatched last April. As I use a trap-nest, I know that these are not their first eggs. I also have some cockerels that have sores on their combs. The sores are dry, sink in a little, and look as if they were covered with a little of the yolk of an egg, but it is not, as they have been kept alone. Please give me some information in regard to them."

Reply: Either of these troubles is liable to appear at any time in any poultry-yard where the conditions are favorable. Hence it is well for all to be informed on these subjects. I hope your questions may lead to helping others, as you have been helped by the answers to theirs. The first trouble you mention is known as prolapsus, or eversion, of the oviduct. The cause is always weakness, or inflammation of the egg organs, especially; no doubt, in this case, the duct that carries the egg to the surface of the body. In order that the egg shall move without carrying the duct with it because of the friction, there must be an abundance of mucus, and the muscles which do the work must be strong and in perfect condition. Good muscles are made by good food and exercise, so it follows that to furnish plenty of scratching material to birds in confinement is a practical means of forestalling such an evil as prolapsus. It is well to consider that there are good reasons back of most of the methods generally advocated, and one who is not posted should follow carefully the recommendations of older workers, even if he does not know the reasons therefor. But I always take especial pains to give reasons, because nine persons out of ten will follow directions more carefully if they know the reasons back of them, and they are, of course, more intelligent in doing it — in other words, it helps their judgment.

I have given a reason for prolapsus; yet the average person will be still in the dark if I do not give reasons back of reasons. That is, we need to know what may cause weakness, dryness, etc., of the egg duct. Inflammation is the chief cause, and inflammation is, perhaps, most likely to be due to the male bird, or to constipation, unless the hen lays an abnormally large egg. Beware of mating a large male with small hens; feed so that the fowls shall not be troubled with constipation, and you will not be likely to have very much trouble along this line. For the same reasons it is desirable that the pullets approaching maturity shall not be annoyed by a lot of young cockerels; hence, the oft-repeated recommendation to separate the sexes after the first six weeks. Doubtless the majority of poultry raisers do not do it, and many of them may think it is nonsense, but all these lay themselves open to having just such difficulties as yours in their poultry-yards. Troubles may come from forcing too hard for eggs, using too much meat and other stimulating stuff. Some have even used ergot and cantharides; then have run to the hen doctor to know why their birds should be in trouble. Do let us all use common sense when we handle living animals, and, if we have not other information, base our judgment, to a degree, on the fact that man, too, is an animal.

Though there is more than one disease with somewhat similar manifestations to those you mention, I think the other trouble is almost sure to be chicken-pox. It is contagious, sometimes very mildly, sometimes quite virulently. So it is better to quarantine and to disinfect. Carbolic ointment is an excellent treatment, with previous washing with strong soap-suds, if you are willing to take that trouble.

"Kindly tell me what is the matter with a hen of mine? She is a Barred Plymouth Rock, full bred. I have sixty of them, and all are healthy but this one. They have a free run. I feed mostly on corn, bran, and oats. When I noticed her first, she acted strangely. She would start and run round and round until she would fall, and she actually laid an egg lying on her back. When she gets over her fit, she sits all drooped looking and doesn't go up on the roost. I notice her droppings are green, mixed with a white sort of stuff."

Reply: It is, in one sense, idle to speculate as to the exact cause of any peculiar difficulty of this type which affects a single individual of a flock. It is possible that no one could correctly diagnose such an individual case; but I think it can safely be said that the affection is a nervous one. And it may be possible to gain a little insight that will help at least to recognize such affections, and not to be unduly alarmed concerning them. The fact that the rest of the flock are all right shows that this is an individual affair, and that conditions of care, feed, etc., are probably normal. It is the hen that is abnormal in some way. Then the question arises: Is the difficulty from without or from within?

So far as my experience with these cases goes, it indicates that they are usually brought on either by excessive heat or by accidents, when from without, and from some digestive abnormality when from within.

Poison, or something excessively irritating to the digestive organs, might induce peculiar behavior resembling what you speak of as a fit. The disease called

HANDLING OF COMMON DISEASES

"limberneck," in which the affected bird is unable to control the head, which droops to the ground, is said by many who have had large experience with it to be the result of stomach irritation brought on by eating maggots. I have seen isolated cases in which I thought it had been brought on by excessive heat. But this, of course, I could hardly prove.

An unsuspected cause may lie in the common accidents of the chicken yard. A case of exaggerated "limberneck" which we lately had will show this. Two surplus males were confined in adjoining coops, built in a series, so that two coops had only a partition between. One of these birds was a late-hatched Wyandotte, quiet, not overvigorous, and not a bird that one would suspect of causing trouble in the hen yards. The other was one of the "watch me" kind, full to overflowing of life and vigor, early hatched, a Leghorn, a little of the Game type, a bird that had whipped everything in sight for several months just for the sake of having it well understood that he was boss. In some way the light partition between the coops was broken and these two birds got together. When found, the meek Wyandotte was just comfortably bloody; the fighter the worst-whipped bird I ever saw, with his body out of control, carriage almost perpendicular, and head thrown backward. By desperate effort, after some hours, he got his head down, and for two or three days it hung limp to the ground. The bird could not eat or drink, and was entirely helpless as to his head and neck. Had we not known of the fight and the character of the two birds, the situation would have been inexplicable. But this gave the key. I have no doubt that the Wyandotte got in a good strong blow somewhere on the spinal column of his enemy and put him out of the fight at once. For about four days I thought he could not recover, and for nearly three weeks he showed some lack of bodily control.

Some one might throw a stone and hit one bird in a way to bring on such an affection. I have seen a four-months' chick get a broken leg just through a clod tossed carelessly to scare him away from mischief. We have even had a new-hatched chick brought in which had some sort of a "fit" every few minutes. I think this was abnormal when hatched. A very faithful sitter suddenly left her eggs. After a day or two it was discovered that she had lost partial control of her head; but instead of circling around she would make erratic dives with her head, first in one direction, then in another. Such affections frighten the birds themselves and make them fear everything that comes near them. I could find no cause for this trouble unless it lay in the fact that she had been let out after being confined for some weeks in a building,

HANDLING OF COMMON DISEASES

and it chanced that the first day she was out was an exceedingly hot one.

At one time we had a hen which behaved exactly the same as the one just noted. Of the cause I never had the least inkling. She improved after a little, but in the subsequent two years that we kept her she would always have a temporary attack if frightened or taken by surprise, this showing plainly enough that the difficulty was purely nervous. I suspect that the Leghorns may be a trifle more inclined to such troubles than the heavier, more phlegmatic breeds.

These cases are interesting to study, but it is scarcely likely that all our study will hinder an occasional abnormality among our flocks. It may be said, I suppose, that perfect physical condition is the best safeguard.

"I would like your opinion of a disease my young chickens have been troubled with. My breed is Silver Hamburgs. They seem bright and hardy for the first five or six weeks, then they droop their wings. They have looseness of the bowels, and in a week or so die. During the sickness their appetite remains about normal. Their feed has been chiefly cracked wheat. They had plenty of grit and were free from lice. It has been the same with both hen and brooder chicks. Will you kindly give cause and remedy?"

Reply: It is never safe to feed any little chicks which belong to the early feathering classes on a diet

T

of one grain only. Of course, cracked wheat is an excellent feed, and wheat is perhaps nearest to a complete feed of any single grain, but the chicks need more than this. If you will listen to the wisdom of a poultry raiser of fifty years' standing regarding the hardships of feathering, you will see, I think, how serious a matter it is to little chicks. He says that it takes many times more nutrition and strength in the hen to grow a coat of feathers than it does to produce eggs. And you know how few hens, even when mature and vigorous, can grow a coat of feathers and lay eggs at the same time. The tiny chick has a small body, a small food capacity, and is using food for growth all the time. The salvation for any animal whose feeders do not understand the value of various grains, etc., is as large a variety as possible, so that the system may have so many more chances to get just the elements that it demands from the food provided.

Probably the most serious lack in the ration you have been using is that it does not contain meat; nor do you mention green food. Both of these are a necessity, especially at the time of feathering. I've seen whole generations, almost, of the most beautiful little active and chipper Leghorn chicks die at just about the period named, from lack of enough variety and enough meat just when this extreme demand comes. Besides the regular feeds, I would keep a dish of mixed granulated grains and bran and some meat and granulated charcoal where all the chicks can have access to it all the time. This gives a chance for life to any that have for any possible reason been hindered from getting a full share of the ration allowed. If you can't get hold of any other green stuff, chop even raw potatoes fine and give once a day for a time, till you can grow a little oats or lettuce, or get some cabbage, which last is the handiest, and, I think, the best you can get, out of the growing season.

"I would like you to tell me what the matter is with my ducks. They are last-spring-hatched ducks, and they were all right until about the middle of September, when I went away and left them with a friend of ours. When I got them back again, about six weeks ago, they were very thin in flesh and looked very rough. I have no idea where they kept them or what they fed them. Before that I fed them cracked corn, kaffir corn, buckwheat, and oats. Since I have got them back I am feeding whole corn, oatmeal, cooked potatoes, and meat scraps from my own table. Sometimes I give them a feeding of bread crumbs and pancakes made of wheat and buckwheat flour, half and half. About ever since I have had them this last time, they seem to have the use of their legs only by spells — a little while at a time. Is this rheumatism, and what shall I do for them? They have a dry place to sleep in, and I give them oat chaff for bedding. It is not a very warm place."

If you had seen, as I have, whole flocks of domestic ducks sleeping the winter through, in thrift and apparent comfort, on a pond covered with ice, except where fed by a cold spring, you would not feel that moderate cold or dampness would injure them seriously. It might be true, however, that a damp house would be worse for them than sleeping on the water. The chief trouble with the legs of fowls appears either as weak legs or rheumatism. Weak legs come usually from overfeeding or giving feed which does not contain enough bone material. This trouble of yours may be rheumatism, though I do not dare to say it is.

I suspect that the whole difficulty began before your neighbors had the ducks, and that you did not notice it. The earlier feed is far from what duck breeders make most use of, as they usually depend on corn for the grain feed, and use much bran with a little meat, corn meal, and sometimes wheat middlings, for the mash. They expect to furnish sand for young ducks to get outside of in some way — more often, perhaps, in the mash itself. Some say a small handful to each quart of feed. One-third of their ration ought to be green food, which is eaten eagerly if cut fine and mixed with the ground stuff. Boiled vegetables are all right for them a part of the time.

Mr. Hallock is one of the largest duck raisers in this country. His ration for breeding ducks is given as follows: four pails of corn meal, two pails of bran, one pail of middlings, one pail of wheat and one of

HANDLING OF COMMON DISEASES

oats, mixed with two bushels of chopped grass or greens or chopped clover, this last dry when green stuff cannot be had. This ration is nearly half green stuff by bulk. If you read "parts" for "pails," you can easily make it of any desired quantity.

"Tell me how to break my hens of eating their eggs. I feed them cracked corn and whole wheat; also scraps from the table. They have ground bone, oyster shells, ashes, lime, and sand. What more can I feed them? If I knew what ones eat them, I could get rid of them, but I never find them in the act, but find the nests wet where they have eaten them. I have never fed my hens the egg-shells, so they did not learn it in that way."

Reply: In order to know how to handle hens in any crisis, it is necessary to have knowledge of their habits of — well, say, temperament. In many points they are just like children: what they are in the habit of doing it is very hard to break them of; and, whether Satan has to do with it or not, there will be mischief of some kind among the fowls if they are idle. Beyond this, there is always a possibility of eggs being broken where the fowls are housed, and thus temptation is thrust upon them. These things being true, it follows that keeping the birds busy is one great step toward keeping them out of mischief; also that the most sweeping reform along the line of inducing good behavior is to so handle them that they will not stay where the eggs are any longer than is necessary. This last point is managed by many by having a laying room and a staying room, so to speak. And if the laying room is less light than the living-room (or scratching shed, or whatever you choose to call it or make it), other things being equal, the fowls will always prefer and seek the lighter room. One movement in the right direction, if two rooms are not available, and it seems necessary to keep the birds confined, is to gather the eggs often. Most of this is preventive, as you see, and, to most people, prevention is far more valuable than cure.

But, the mischief being done, if the fowls must be kept housed among the nests, there is this that may be done: they may be kept scratching about all the time; they may be fed such things that the egg-shells will become too strong to crush easily, and if extreme care be taken to furnish them shell, grit, and bone, the shells may be made so hard that they will find it difficult to break them. Feeding green stuff in sufficient quantity to overcome the abnormal appetites that are liable to seize a grazing animal cut off from grass will be another help. It is, of course, good sense to turn the nests to face away from the light and place them high enough so that the fowls cannot see into them all the time, and arrange them so that they cannot stand in front of them much. All these bad conditions together are found in many poultry-houses, and, as the hen appreciates good eggs even better than her owner does, the wonder really is that more of them do not become pronounced egg eaters. If all else fails, the nests must be fitted with false bottoms of cloth, so that the egg, as laid, will pass through a slit and be beyond the hen's reach. Open-air life is the best insurance against egg eating.

THE INDIAN RUNNER DUCK

Its Appearance and Habits — Its Utility Value — Its Claims to Recognition in America

WHILE the Indian Runner Duck fails to reach the American ideal in being a general-purpose fowl, it is so strong in its specialty of egg production, and so well adapted to fill a special place in the flesh market as well, that it is crowding its way forward to an unprecedented degree. Though these birds have been known in the British Isles for many years, it is now scarcely five years since the first inquiry regarding them, showing the awakening of popular interest, reached my desk. At that time a few scattering articles from British writers in our poultry publications about comprised the literature concerning them in America. They were in our 1905 Standard, but I had not then seen one in the shows, though I do not say that none had been shown.

No topic concerning which I have ever written for the help of poultry keepers has proved so interesting and so popular as this. Letter after letter comes to me, asking questions, suggesting articles, or backing up statements previously made, by experiences from the writers' own yards. Yet I do not know of another breed of ducks which has been so persistently decried in some quarters, chiefly by those who knew little about them, and who have allowed a previously formed prejudice to warp their judgment.

The Indian Runner has changed all this by sheer force of merit. It has forced its way all over the country, and from all quarters comes the same story regarding its value. It is rapidly making a history, and hence a literature of its own, through its performance upon American soil, and I know of no other fowls concerning which growers thereof are so unitedly enthusiastic. Whatever may be said to the detriment of fanciers as a class, I think it is wholly true that they are filled with satisfaction when they discover a variety which can "make good," — as they like to say, — one for which they do not have to apologize at some point. The best exponent of such a variety at present before the public is the Indian Runner duck. The plant, the flower, the fowl, that can best illustrate the word "satisfactory" is a most excellent one to handle. It is at the head of this class that the Indian Runner stands.

The American Standard, as at present operative, shows a bird long in body, comparatively thin from breast to back, looking as if flattened, with a thin, long neck, an upright carriage, a bill running in a straight line with the top of the head, and a main tail on the same line with the flat back. Its peculiarity of appearance is described as "racy-looking" — a word which neither Webster nor the Standard glossary defines. Probably the most illuminating phrase in the Standard description is, "resembling the penguin in form." And yet, as one looks at the long, slim neck of the Runner, with its trimness of feather, and notes the short, thick neck and head appendages of the penguin, he will begin to think that the Standard makers did not express themselves quite accurately, and that they probably meant "resembling the penguin in *carriage*." A common expression for this carriage would be that the bird "stands on end." As pictured, the angle of the body carriage is about forty-five degrees, which is very upright, as compared with most domesticated fowls. The present (1909) colors are light fawn and white, in the preferred form, though gray and white has been allowed as an alternative combination. I think breeders have already been throwing gray out as much as possible, — it surely was an error to allow it, — and the next Standard will not allow gray and will omit the adjective, making the colors fawn and white. These are almost equally divided, the head and anterior part of the body being fawn, the neck and the posterior part white. On the back, the darker color runs toward the tail in a point so as to form a very true heart shape low on the back, though it cuts straight around the neck in a clean line of demarcation. The only break in the contour line of the back, when the characteristic carriage is assumed, is made by the curled feathers in the tail of the matured drake.

Contrary to the idea gained by the usual references to the upright carriage, the Indian Runners, as seen in America, do not show this carriage notably, except when in fear, or when running. But it is the one point on which judges will be likely to insist, as far as possible, since it is the most distinctive feature of the appearance, and gives the breed its name.

The Runners are, essentially, gamy birds. They have not forgotten the days behind them, when every hillock might hide an enemy, and the males, especially, are constantly on the lookout for danger, and announce it many times, in domestication, when none is near. This causes stampedes, frequently, and makes the entire flock grow more wild. For this reason it is wise to reduce the number of males to the required breeding numbers as early as possible.

It may be considered practically impossible, at the present stage, to overcome, wholly, the timidity of the Runner. But a great deal can be done by taking special pains to tame them when young, to avoid movements which frighten any nervous bird, and to remove the males that are unnecessary as soon as may be. The

Indian Runner has been called the Leghorn of the duck family, and those who dislike Leghorns because (usually) they are themselves of a nervous temperament, will not be wholly successful in handling the Runners.

Although nervous and restless at times, the Runner rests completely when it does rest, and accustoms itself to confinement rather more easily, I think, than most breeds of fowls. It will thrive and lay exceedingly well in confinement and with only water enough for drinking, so that not only can it be raised profitably on farms which have natural springs or streams, but on dry-land places as well. The birds are classed as non-sitters, but an occasional one will manifest a desire to raise a family. The males are sometimes troublesome, and it is altogether wise to keep their number as low as may be. This is, indeed, a broad wisdom, in dealing with all fowls, for a lot of unproductive and troublesome males is a decided detriment to any flock, and a source of loss in money.

As to the value of the Indian Runner, there is no dissenting voice, as far as my knowledge goes, among those who have given it a fair trial. Indeed, there is enthusiasm galore. A woman who characterized her experience with the Runners as "a profitable and a happy one," stated that her birds laid almost continuously from January to July, and, after the moult, began again early in October. From three stock ducks she raised, the first year, one hundred and fifty young, and she stated that they weighed three pounds each at nine weeks old. "An investment of thirteen dollars," she says, "has given me a flock valued at three hundred dollars. Is it any wonder I am an ardent admirer of the Indian Runner duck? She is a friend to the poultryman, and a veritable gold-mine to the farmer. He can go out every morning and bring in his basketful of eggs, which bring from five to ten cents a dozen more in the market than chicken eggs."

A New York man, having a small flock under the first trial, states that his birds, hatched in June, began laying in January, and from February 15 to July 24, in one stretch, they laid an average of one hundred and thirty-three eggs. This was one hundred and fifty-nine days. As a table fowl he considers that they surpass anything within his knowledge. He, as well as many others, mentions the small amount of food eaten. One eulogistic woman writes that the food needed by a Pekin duck for one day will last a Runner a week! The Runner weighs four and one-half pounds, when not above Standard weight, and might naturally be allowed one-half the feed of a Pekin. They *are* small eaters, but I think the above may have been "a guess." The question of raising the Standard weight of the Runner has been discussed not a little by interested breeders, but the feeling is very general that raising the weight would almost surely result in lessening the constant-laying tendency. In shipping a coop of young males recently, I weighed several. None went below four and three-fourths pounds, one weighing five. This shows that we can raise the weight if we desire to do so.

Some time ago a correspondent sent me a record for the year, for free-range stock. The average number laid was one hundred and eighty-five. This is by no means as high as is claimed for these ducks, two hundred being, apparently, regarded as quite easily reached, while two hundred and fifty is freely claimed. At any rate, a common expression among those testing them, is, "They beat all I ever saw, for laying."

"They go ahead of anything I ever had in feathers," said one of my correspondents.

As a fairly conservative statement I would like to quote from one who had, at the time of making this public estimate, bred the birds for fifteen years. "I have kept records of the different pens, and in many cases the average has been up to one hundred and ninety eggs in the year. The best individual layers do not lay over two hundred eggs, but the average for a good strain of layers can be safely set at one hundred and eighty per annum. The [weight] average is twenty-one pounds to one hundred and twenty eggs, so that a duck which weighs four and one-half pounds and lays one hundred and eighty eggs produces exactly seven times her own weight." The same breeder (writing for the British Isles) says: "This duck is at her best, so far as egg production is concerned, during October, November, December, and January, and lays until the end of July."

From far New Zealand comes corroboration of this, in the statement that the Runners "lay when eggs are dear, and when most of the hens go on a strike." The author of it, claiming to be the manager of "the largest poultry plant in New Zealand," avers that he has had an average production, from a flock of twenty, of two hundred and thirty-four eggs. With others added, the bunch averaged, in the second year, one hundred and eighty-four, and, being made up to sixty, the lot averaged, the third year, one hundred and eighty-three. Two ducks in a pen apart laid two hundred and forty-two each, within a year. The feed usual with this flock was bran and middlings, 5 per cent of meat meal, whole corn at night, and as much green feed as of bran. The photograph of breeding pen published by this manager shows birds so like our own that they might be mistaken for them. He states that he reared, in a single season, eight hundred and forty-four out of nine hundred hatched; this is a strong point, the loss always being small when the care is reasonably good.

I took up these birds solely to prove them, and have found them the most interesting birds I have ever handled, with the exception, possibly, of Embden geese. I have found them more reliable layers than any of the fifteen varieties of hens that I have had at various times during a long period of years. This includes most of the American varieties, Minorcas, and both White and Brown Leghorns. The ducks have never been later than February in taking up the spring task, and, when early hatched, they have never laid at later than five months, one lot of March-hatched laying at four and one-half months. Others have claimed that they would lay at five months, no matter when hatched. Ours have not reached this record, if hatched late in the summer. But no one can be too arbitrary in making statements as to what is impossible for these birds, as ducks will stand pretty heavy meat feeding, and the amount of meat fed, other things being equal, will probably decide the amount of the egg product.

During the past season I visited the yards of a most enthusiastic breeder of the Runners, and found over one hundred and twenty ducks of fine quality and in thorough vigor. She told me that she received at least eight cents above the going price for hen eggs for all her Runner eggs. Why not, since they weigh one-half more, on the average? Later on, I received a letter from her containing this bit of record: "My free-range flock averaged 75 per cent for July, 50 per cent for August, and 34 per cent for September. They are about through moulting, and laying as well or better than last month." This was written October 11. I regard this breeder's statements as absolutely reliable. And I do not know of "anything in feathers," — to quote my other friend, — aside from the Runners, that will equal this record.

In August, 1909, while attending a large gathering of the men who are influencing most strongly the poultry industry in the United States, I made it a part of my business to talk with some of the foremost breeders of Indian Runners regarding this breed. They regarded its future as very bright. One of them who has not even made Runners his specialty, told me he filled one order for five thousand eggs last season. Yet it seems to be the feeling that the demand of the immediate future will be greater than the supply.

I was in an interested conversation with a keen man of affairs in New York City. Our topic was the three-hundred-egg hen. This gentleman said: "I do not believe it is possible for hens, as they are now constituted, to lay three hundred eggs in a year. We may have, in the future, something with feathers, maybe looking something like a hen, that can do it." I think we may see the answer to this semi-prophecy in the Indian Runner duck!

I have looked over a stack of copies of an important poultry magazine, the numbers covering several years more or less completely, and beginning with

1904. In the period covered thus I found but two articles on Indian Runners, and noted no other references to this breed. Now one can scarcely take up an issue without finding some reference to the breed. This shows how strongly and quickly its popularity has come upon it. In a word, it seems that the Indian Runner *commands* its following.

It is a fact well known to those who have taken any interest in the subject that duck eggs have not been offered in the largest markets, in earlier years, after midsummer, and it has been taken for granted by some that they were not desired after this date. Inquiry among commission men has brought out the statement that they have not been quoted simply because they could not be obtained in sufficient numbers to warrant it. With the spread of the Indian Runner there is the best of reasons to believe that quotations will appear later, and throughout the season. Restaurants, bakeries, and egg-drink establishments will serve their own interests in a high degree by calling for these duck eggs, as, even at an advance over the price of hen eggs, they will be economical. The fact that they make very superior custards, and most delicious omelets, by using a little more liquid than hen eggs will bear, makes for economy and gastronomical pleasure as well. A physician in one of the large Eastern cities has prescribed them for his patients as being better than hen eggs, and they are constantly increasing in public regard.

While feathers are not the item they once were among the necessities of life, there is still sufficient demand to hold the price above what middle-class people can afford to pay. Since duck feathers are recognized as being far superior to hen feathers, the additional income to be had from the sale of feathers is by no means to be overlooked. I do not now recall, in fact, any other farm product which will enable the grower to turn his money so many times in a season, and so profitably, as this duck. This quick turning of money with a profit is the key to business success — a fact so well known as to be an axiom with retailers.

A special market for the flesh is likely to open, it is thought, along the line of supplying the deficiency in game birds. The bird is small, the flesh is gamy, and very toothsome indeed, and the market for real game birds is turning more and more to other birds to supply its demand. The guinea fowl, though it has entered this field, and has risen in price of late, cannot, in the nature of things, furnish so continuous and large a supply as the Indian Runner duck can easily give. Thus there seems to be no reason to expect other than that the Runner will enter a field so large, and which it is so well fitted to fill.

With the idea of feeling the market (or perhaps posting it in advance) with regard to the Indian Runner duck, I wrote to one of the oldest and most reliable commission firms in New York City. I described the duck, in a general way, and asked if it had appeared in the market, and what the chances might be as to its taking a "game" position. The reply stated that they were not yet familiar with this duck, by name, although they might have sold it, unknowing. The statement was absolute that almost any good bird would sell well, the chief point being that they must be fat and of good flavor. As the Indian Runners will be fully responsible for the good flavor, it remains only for the owners to see that they are well fattened. The market will not fail. As I write the closing words of this study, word comes that, in our local market, chickens took an advance jump of five cents during the holiday week, selling, dressed, at twenty-eight cents per pound. Also from New York market comes the report of carload lots of turkeys selling at twenty-four cents, which means, probably, above thirty cents to the consumer. The American hen and her kindred are commanding the market as never before! To use the words of our Secretary of Agriculture, "This industry has advanced at such a rapid rate that no arithmetic can keep up with it."

INDEX

Abnormalities, occasional, 273.
Accidents, common, 240.
Acid, carbolic, 256.
Advance sheets, Government report, 1.
Arithmetic, behind poultry industry, 292.

Beauty *vs.* utility, 117.
Billion-dollar hen, the coming, 112.
Birds, coming: Orpingtons, White Rocks, 35.
Blockiness, craze for, 95.
Board of Trade, enterprising, 147.
Bone, green, may cause diarrhœa, 265.
Breeds, American, Buff Orpington, notable rival, 52.
Breeds best, in Australian tests, 62.
  best paying, 10.
  color domination, 13.
  competing, abroad, 56.
  early, Javas, Dominiques, 20.
  Standard, American class, 13.
Broilers, 137.
  Guinea, 4.
Bulletins, value of, 168.
Buyers and reasons, 137.

Cantharides, blistering skin, 172.
Chicken-pox, cause, treatment of 254, 269.
Chicks:
  affected by feed, 191.
  best time for hatching, 196.
  beware of musty grain for, 194.
  box raising, 199.

Chicks, distributing mash, 193.
  doubling up, 204.
  dressed, advance in market of, 292.
  extravagances, 195.
  farm, confining, 204.
  February and March, 197.
  feeding, character of mash, 193.
    continuous supplies, 192.
  first cost of, 140.
  give sunshine early, 189.
  good eggs necessary to, 198.
  good tendencies of, 207.
  lack of water fatal to, 198.
  late, 197, 200.
  meat for precocious feathering, 194.
  shelter for, 201.
  summer handicaps of, 198.
  the "contingent fund," 199.
  too close coops ruin, 202.
  trampling each other, 203.
  varied diet a necessity to, 274.
  vermin on, 205.
  weakly, a prey to diseases, 131, 132.
  weaning coops for, 202.
  weather changes affecting, 188, 189, 190.
Cockerels, many, cause loss, 135.
  sell early, 139.
Colds, neglected, 258.
  routing, 253.
Color, chicks, various breeds, 184, 185, 186.
  buff, fading, 31.
    vs. white, 31, 33.
  ducklings, Runner, 187.
  foreign, in Brown Leghorns, 187.

293

INDEX

Comb, indicating condition, 205.
Condition, indicators of, 206, 208.
   laying, handling to judge, 265.
   of droppings important, 209.
Constitution, good, vital, 130.
Coop "A," corners of, 243.
   weaning, 202, 230.
Coops, small cornerless, 203.
   stuffy, 201.
   summer, 202.
   too small, 201.
   unsheltered, 190.
   with roosts, 204.
Cornish, White Laced Red, 84.
Cracks, lead to distempers, 261.
Crippling, temporary, due to male, 266.
Culling, importance of, 130.

Diarrhœa due to rich mash, 266.
Disease, due to faulty conditions, 249.
   foresight concerning, 251.
   nervous, 270.
   nervous, causes of, 270, 272.
Diseases, summer, 249.
   nervous, 250.
Distempers, 257.
Dominique, American:
   hardiness of, 21.
   increase in favor, 21.
   precedence of, 21.
   type of, 21.
   why behind Rock, 21.
Doors, front, to each pen, 235.
   screen, 236.
Drugs:
   disinfectant, 177.
   fatal forcing with, 171.
   in condition powders, experiments, 179.
   in drinking water, 178.
   stimulant, in egg foods, 176.
Ducks:
   good, sell well, 292.
   Indian Runner,
      carriage, 282.
      colors, 282.
      decried, praised, 281.

Ducks:
   Indian Runner,
      early, reliable layers, 288.
      eggs, bonus for, 288.
      enthusiasm of breeders for, 284.
      flesh, special market filler, 291.
      in New Zealand, 287.
      laying records, 285, 286, 288.
      merit of, 281.
      quotations on, 290.
      Standard illustration, 281.
      superior, economical, 290.
      thrift in confinement, 284.
      timidity of, 283.
   unthrift, lameness, 276.
   vegetable food necessary, 277.

Egg:
   laying, averages with large numbers, 159.
      depends on digestion, 158.
      general averages of, 159.
      low average, 163.
      phenomenal flock averages, 160.
   production, claimed records of, 160, 162, 163.
Egg farms, system of branch 117.
Eggs:
   breeds vary as to fertility, 152.
   brown, most common, 143.
   chilled, supposed infertile, 153.
   dark brown, unattractive, 142.
      not richer, 142.
   fertility, care improves, 153.
   hatching qualities, 148.
   hens eating, 277.
   infertiles, for chicks, 149.
   judging breed by, 141.
   knowing fertility of, 152.
   losses, gains, though infertiles, 149, 151.
   of various breeds, 144.
   perfect, what makes, 151.
   source of income, 8.
   shape of, 143.
   Standard for, 142.
Error, greatest, on farms, 150.

INDEX

Expenses, fanciers', 116.
    make birds pay, 248.
Experiment, fresh air, drastic, 228.
Experimenting, coöperative, 68.

Failures, cause of, 134.
Fancier:
    builds foundations, 106.
    gives farmer uniform standard, 107.
    must be good business man, 103.
    must be practical, 104.
    the, at work, 101.
    works for uniformity, 105.
Fancy and utility, mutual supporters, 99.
Farms, small, near towns scarce, 127.
Faults, in feeding, 210.
Feathering, chick, critical period, 274.
Feed, surplus provision, 275.
Feeding, lack of judgment in, 263.
Feeds, combining, 264.
    forcing, 167.
Flies, blister, cause inflammation, 172.
Foods, stimulating power o , 133.
Fowl, market, best form for, 121.
Fowls, condition of, 205.
    market, "fricassee," 123.
Front, cloth, compromise on, 232, 235.
    movable, 231.
    muslin, double, 229.
    too warm, 230.

Game, Government protection of, 5.
    supply affects poultry market, 3.
Gapes, prevention of, 252.
    treatment for, 252.
Gauge, moisture, Government patents, 2.
Guinea, 4, 9, 291.

Habits, hens', to be watched, 134.
Health, selection for, 206.
Hen, American, commands market, 292.
    distrust of, 127.

Hen:
    profitable, must be consistent worker, 129.
Hens, egg-eating, keep busy, 278.
Hiring, impracticable, 246.
House, damp, cause of, 219.
    double boarding, 218.
    floor space, 217.
    for fifty hens, 216.
    fresh air, 225.
    "improved," 229.
    needs drainage, 217, 220.
    sanitation of, imperative, 251.
    Tolman, approach toward, 228, 237.
        evolution of, 227.
        good points of, 226.
    ventilation of, helps warmth, 225.
Houses, "A" style, rather common, 212.
    cement-walled, 221, 223.
    poultry, unusual, 211, 213.

Ideals, varying, 38.
Inexperience, 243, 247.
Inflammations, oviduct, causes of, 268.
Inquiry, letters, actual, of, 30, 171, 175, 216, 220, 224, 262, 267, 273, 275, 277.
Instruction, poultry, public, 246.
Investment, careful, 125.

Java, Black, hardy, 21.
    of Asiatic type, 21.
    Professor Watson's estimate of, 21.
    superior, 20.

Laying, strong muscles necessary to, 268.
Learning, cost of, 243.
Leghorns, Rose Comb Brown:
    as fancy fowls, 64.
    breeders of, enthusiastic, 65.
    eggs, number and size, 64.
    profit, yearly, $5.27, 60.
    quality of, 63, 67.
    record of, 60.
    Special Government notice of, 61.

Leghorns, Rose Comb Brown:
    status of, 61.
    tested with other breeds, 60.
    weight, 63.
    "world-beaters," 67.
  White, high standing of, 61.
Legs, scaly, treating, 252.
Limberneck, causes of, 271.

Market, catching right, 136.
Markets, demands of American, 120.
Meat, gamy, domestic producers of, 9.
Meats, investigation concerning, 7.
Medicines, 259.
Mischief, due to feed and handling, 278.
Money, put into poultry, 124, 129.
  sunk, 126.
Mothers:
  brooders make inferior, 183.
  faults as: Rocks, Buckeyes, Reds, Columbians, 181, 182.
  nearly perfect, White Wyandottes, 182.
  superior, American breeds are, 180.
  what constitutes good, 181.

Netting, wire, diamond mesh, 239.

Ointments, disinfectant, 252.
Orpington, Buff, no Orpington blood, 53.
  Black, egg average, in competition, 56.
    leads its breed, 56.
  Buff, American ideals, 54.
    egg average, in test, 56.
    inferior to Black, in egg production, 56.
    no Orpington blood, 53.
    numbers recently shown, 55, 58.
    origin of, 53.
    popularity, England, America, 56.
    seemingly making good, 55.
Outcrossing, 37.
Oviduct, eversion of, 267.

Partitions, cloth, 234.
Popularity, in general, 34.
Poultry, farm, 1899 values, 112.
  Government interest in, 2.
  pays, proof that, 127.
  rank, probable, 1910, 2.
Prices:
  advance of, 8.
    eggs, sympathetic, 8.
  high, for fancy eggs, 115.
  poultry vs. grain, 8.
  wholesale, average, advancing, 113.
Problems, breeding, 30.
Products, farm, rising prices of, 8.
  value of, 1.
Profit, margin of, 138.
  other meats, 8.
  winter, depends on handling, 119.
Profits, price drop comes out of, 122.
Promises, deceitful, 167, 169.
Pullets, quarter separately, 135.

Questions, beginners', 238.

Rations, elements lacking in, 274.
Records, high, dangerous, 165.
  private, unreliable, 170.
Red, Buckeye:
  a woman's successful work, 46.
  color standard, 48.
  confused with Rhode Island Reds, 44.
  faults, 49.
  first base, Plymouth Rock, 46.
  flesh fine grained, 46.
  good for table, 48.
  hardy, 48.
  origin and name, 45.
  originator's statement, 46.
  richness of color, 50.
Red, Rhode Island:
  advancement rapid, 75.
  color consistency, 41.
  defects, 75, 76.
  hard to breed to Standard, 37, 40.
  origin, 38, 40.
  Rose Comb, once "American," 43.

INDEX

Red, Rhode Island:
    shape inconsistency, 40.
    Standard requirements, 40.
    under color, 42.
    William Tripp's work with, 39.

Rock, Barred Plymouth:
    color terms, 14, 15.
    difficulty in breeding, 12.
    foreign color in, 11.
    laying capacity, 16.
    low egg average, 17.
    most widely bred, 16.
    on farms, 15.
    scale of points, 13.
    Standard weight, 14.
    Buff, as a layer, 18.
      fattens too easily, 19.
      valuable, 18.
    White, bidding for favor, 76.
      a "find," 17.
      a safe choice, 17.
      "coming fowl," 17.
      critical estimate of, 18.
      mammoth specimen, 20.

Roosts warm, at night, 231.

Roup:
    applying liquids to mouth, 174.
    attributed to warm houses, 233.
    duct stoppage causing lumps, 173.
    use of kerosene, 175.

Season, selling, adapting stock to, 122.
Shade, abundance of, a necessity, 156.
Shelter, warmest, 232.
Standard of Perfection:
    a necessity to breeders, 87, 91.
    basis of laws, 94.
    color plates, 30.
    copyrighted, 93.
    difficulties in making, 96.
    exact coincidence with, 36.
    glossary of, 93.
    keeps up weights, 100.
    law for breeders, 92.
    making breed distinctiveness clear, 29.
    mental standard, auxiliary to, 88.

Standard of Perfection:
    modifies foreign types, 94.
    revised, 1909, 86.
Stations, Experiment, results of work, 170.
Statistics, expected to be fuller, 111.
Storage, cold:
    chickens, danger in holding, 146.
    egg-producers availing selves of, 147.
    in New Zealand, 145.
    investigation of, 146.
Stories, impossible, 164, 167, 169.

Team work in poultry, 140.
Treatment, therapeutic, failures in, 259.
Trees, as roosts, 261.
Type, egg, 165.
    general, 16.
    photographs deny common idea of, 166.

Utility, three chief points of, 118.

Varieties:
    "any other," 79.
    beauty types, American, 84.
    bleaching white, 83.
    color deciding rank, 74.
    Columbian to lead, 83.
    competing Wyandottes and Rocks, 81.
    Government introduction of, 71.
    interest in fancy, 82.
    new, enough now, 70.
    popular in our shows, 79.
      competing Wyandottes and Rocks, 81.
      mostly of similar type, 80.
    pushing by advertising, 59.
    show reports on, 82.
Vermin, marauding, 242.

Water, continuous supply imperative, 241.
    swales dangerous, 242.

Wealth, poultry, increasing, 112.
Winds, cause of colds, 260.
Winners, great, and layers, not all white, 33.
Wyandotte:
  Black, not widely bred, 27.
  Buff, appearance, 28.
    eventual rank, 28.
    laying capacity, 27.
    leads Wyandottes in laying, 59.
    vs. Buff Orpington, 59.
  Columbian, good winter layer, 65.
    present popularity, 28.
    prophecies concerning, 28.
    rapid advancement, 28.
    selection must improve, 29.
    vs. Barred Rock, 29.

Wyandotte:
  Golden, a beauty bird, 26.
  origin of, 23.
  original type of, 24.
  Partridge, difficult to breed, 26.
  Silver, competition honors, 25, 26.
    honor due to, 24.
    most valuable descendant of, 24.
    refused Standard admittance, 23.
    the breed progenitor, 24.
  White, a bird of curves, 25.
    egg record, 25.
    is improvement possible? 78.
    near American ideal, 76.
    shortening body of, 25.

Yankee, on trial, 110.